Enter the Animal

ANIMAL PUBLICS

Melissa Boyde, Fiona Probyn-Rapsey & Yvette Watt, Series Editors

The Animal Publics series publishes new interdisciplinary research in animal studies. Taking inspiration from the varied and changing ways that humans and non-human animals interact, it investigates how animal life becomes public: attended to, listened to, made visible, included, and transformed.

Enter the Animal: Cross-species Perspectives on Grief and Spirituality

Teya Brooks Pribac

SYDNEY UNIVERSITY PRESS

First published by Sydney University Press
© Teya Brooks Pribac 2021
© Sydney University Press 2021

Reproduction and communication for other purposes

Except as permitted under the Act, no part of this edition may be reproduced, stored in a retrieval system, or communicated in any form or by any means without prior written permission. All requests for reproduction or communication should be made to Sydney University Press at the address below:

Sydney University Press
Fisher Library F03
University of Sydney NSW 2006
AUSTRALIA
sup.info@sydney.edu.au
sydneyuniversitypress.com.au

A catalogue record for this book is available from the National Library of Australia.

NATIONAL
LIBRARY
OF AUSTRALIA

ISBN 9781743327395 paperback
ISBN 9781743327401 epub
ISBN 9781743327425 mobi
ISBN 9781743327432 pdf

Cover image: Teya Brooks Pribac
Cover design: Miguel Yamin
Painting of Charlie: Teya Brooks Pribac
Quoted material from *The Outermost House: a Year of Life on the Great Beach of Cape Cod* by Henry Beston. Copyright © 1928, 1949, 1956 by Henry Beston, Copyright © 1977 by Elizabeth C. Beston. Reprinted by permission of Henry Holt and Company. All rights reserved. Pushkin Press is the publisher of this work in the territories UK & Commonwealth (excluding Canada) and has also provided permission for use of the quoted material by Henry Beston in this book.

Contents

Foreword

As a Freudian analyst and director of the Freud Archives I was given access to a large treasure trove of unpublished Freud letters that seemed to show that the history of child sexual abuse we had been given (namely, that it was mainly a female fantasy) was false, and that child sexual abuse was real, serious, and everywhere. Freudians did not like that. I was fired from my position and lost my licence to practice psychoanalysis. I had also by that stage given up a full professorship of Sanskrit at the University of Toronto. So now what should I do with the rest of my life?

This was in 1980. I was living with the great law scholar, Catherine MacKinnon, and she said: 'Start by reading about what you love.' That was easy. Even though I was no longer an analyst, I was fascinated by emotions, but having grown disenchanted with the human animal, I decided I wanted to explore the inner world of other animals. I wanted to learn everything there was to know about animal emotions in general. Easier thought than accomplished.

I turned to the author of *The Question of Animal Awareness*, Donald R. Griffin, who had been a distinguished professor of biology at Harvard, and had discovered bat sonar. I called him, and he was absolutely lovely, but he could not tell me much: 'Jeff, if you want to read about animal emotions, after you have read Darwin, you will have to write about it yourself. There is practically nothing written about it. I got into enough trouble with scientists by suggesting that animals were conscious and aware. I would have been run out of the university had I suggested they had emotions like our own, but I am sure they do.'

That was all the encouragement I needed. I began doing the research and the writing for a book that became *When Elephants Weep: The Emotional Lives of Animals*. I have never had more fun doing research for a book than that one: everyone had a convincing story to tell me about animal emotions, and people bought the book because, I think, they wanted validation for their own feelings on this matter. The book simply stated what was obvious to anyone who had ever lived with a dog or cat or bird, or really, any other nonhuman animal. *Of course* they had feelings. Who could doubt that?

But some did doubt. I was accused of not being a scientist (true) and of committing a major sin: anthropomorphising nonhuman animals, that is, attributing to them characteristics they could not possibly possess (by fiat?). I was projecting my own feelings onto them. I did not believe that for one moment, but as it turned out, the story of animal emotion that we were given – that was forced onto us, that people had to accept if they wanted to save jobs and intellectual reputation – was equally false. Despite being passed as a fantasy of the weak-mind, like child abuse had, animal emotions were also real, serious and everywhere.

Thirty years is a long time. Attitudes have changed. Books on animal emotions are growing more numerous and interesting. Two of the more recent gems that have touched me are by scientists Frans de Waal and Carl Safina. Others too – non-scientists – are opening up to the idea that nonhuman animals not only have emotional lives comparable to our own but also that it is possible (I go further and say it is certain) that they experience some emotions more powerfully than we do. Who can equal the *joie de vivre* that a dog is capable of? Or the level of contentment that cats appear to reach?

What about more complex emotions? That is the territory Teya Brooks Pribac takes us to in this book, aptly titled *Enter the Animal*.

To begin with we learn that 'complex' is a loaded term. Just because we do not understand something, it does not mean that it is more complex compared to something that we understand or think we do. Recent scientific advances have both filled many gaps in our knowledge as well as exposed errors that we considered as truths. The other problem with 'complexity' is that as soon as a phenomenon is labelled as 'complex', nonhuman animals are automatically excluded from the circle of consideration because we have been conditioned to think of them as anything but complex, as simple, basic, unsophisticated. And how wrong we were!

We tend to get lost in our interpretative world and forget that our body continues to live in this other world, touches things and gets touched by them, communicates in ways that we often do not even notice or recognise as meaningful experienced communication. But when we start peeling off those meandering layers of anthropocentric quests and solutions we discover a dimension that is more immediate and grounded in relations, and these relations with significant others – be it other humans, other animals and the rest of the world – enable us to survive, to live, to be who we are and to feel the way we feel. Complex? For sure. Exclusively human? Absolutely not.

In this book Brooks Pribac takes us on a journey deep into the animal body/mind. She began her research, as I did, by wanting to understand something. Then there was reading, thinking, experiencing – in fact, all along she has also been involved hands-on with various animals, some of whom we meet in the book. As she begins to understand, she brings the reader along with her, because she has to be clear in her mind exactly what it is that she is seeing and seeking to understand. That is a quality you rarely find in a book as scientifically rigorous as this one. No matter the angle you are approaching this from, this book offers a unique, concise presentation of the material. It is clear, and easy to read, and easy, as well, to understand. You are treated to the writer thinking out loud as she writes. Whether you are a scholar in the broad area of animal studies, a student embarking upon animal-related research or simply a reader interested in all matters animal, this is an essential book, which will help you understand three fundamental points: where we are currently, how we got here, and where to go next.

— Jeffrey Moussaieff Masson, PhD

Charlie by Teya Brooks Pribac

Introduction

In the final stages of the editing of this manuscript Charlie died. He had irreversible cancer. He died moderately peacefully, at home, in the presence of his loved ones. Thirteen and a half years earlier, Charlie, a black and white fox terrier cross, had been rescued from death row, completely traumatised. It took time, patience and love: Charlie was on the mend. He was always considered an equal member of the family unit and had a good life overall; nevertheless following his death remorse set in. I regretted all the walks we had not taken but could have, all the beach trips we had not done (he loved water), and many other things, including the canine friends he lost when humans and their dogs moved towns and cities for human convenience. Charlie's story, with new ends and new beginnings. Could I have done better?

People sometimes wonder what the function of grief is in evolutionary terms. How does it increase fitness? Physical pain, for instance, is very helpful: it is a signal that something is harmful and the organism needs to attend to it. What about grief? I do not have an evolutionary, or revolutionary, answer but Charlie's passing made me even more aware of the importance of staying present, of being – body and mind – with the loved ones while they are still with me. When they leave, memories are all that is left. They had better be good ones.

Charlie's passing reminded me of another dog, also black and white with many shades of grey and a story with probably as many ends and beginnings. He was my best friend in my early childhood. His name was Bobbie – the Bobbie mentioned later in this book was named after him. He 'belonged' to the neighbours and, allegedly, saved my life twice. He was very bright. 'So human,' everyone would say. 'Speech is the only thing he lacks [to be fully human],' they would add. Despite all this Bobbie had to live outside and had other restrictions imposed upon him because ultimately he was an animal – not a human animal, just animal. The story unfolding on an experiential, immediate and intimate level was subdued by the other story – the official, 'intelligent' one – and the world remained divided.

Historically, a substantial amount of philosophical discourse, particularly in the Western tradition, has centred on attempts to articulate a framework of defining features that would capture the presumably distinct nature of the human. '[S]uspended between a celestial and a terrestrial nature,' Giorgio Agamben notes, the human's being, as it constructs itself, is 'always less and more than himself.'[1] During this quest for essences and meanings, the human ontological fabric has become significantly defined by the human/nonhuman animal binary. Taken for granted, Aaron S. Gross observes, the binary enables the otherwise rigorous contemporary scholar the freedom to uncritically invoke it for support.[2] In order to secure this extremely vulnerable mode of being, separatists have over time built a domino castle of 'exclusively human' attributes only to see it progressively collapse as knowledge and understanding of humans and other animals rapidly increases. Tool use, for instance, was believed absent in other animals' worlds until a variety of nonhuman animals, including octopi from the invertebrate group, were observed operating various objects not only to acquire food, but also for other purposes, such as averting predators as in the case of the makeshift whistles orangutans assemble using leaves pulled off a twig.[3]

Reason was another attribute considered uniquely human, with the human as *rational* animal standing in juxtaposition to all other 'instinct-driven' animals. Increased understanding of cognition and

---

1   Agamben 2004, 29.
2   Gross 2015, 91.
3   Hardus et al. 2009.

emotion, however, has forced the rational animal into humbleness again. The highly praised reason turned out to be far more influenced by emotions (traditionally held in lower regard) than previously believed, but not just that: emotions were found to play an essential, positive role in normative functioning, including in the reasoning process. This proposition was met with scepticism when neuroscientist Antonio Damasio[4] first proposed it but has since been largely embraced. Convergent scientific evidence also shows that the processing of raw data and concept formation (whereby 'concept' is defined as 'grouping of "object" attributes') in the brains of human and nonhuman animals follow the same principles.[5] The contents of the minds and relative associative combinations of course vary across species and individuals within species. However, it is becoming increasingly clear that the differences between humans and other animals are less significant than the similarities, that other animals are far from being instinctual automatons, and that just like human lives, other animals' lives are replete with thinking, evaluations and decision-making, informed by various species and individual specificities as well as their developmental and living contexts.

Entering into the debate opened by these and similar investigations, the present volume examines the underlying commonalities that enable the experience of grief and spirituality as an embodied practice in both human and nonhuman animals. Considerations of both phenomena in nonhuman animals continue to be tainted by anthropocentric philosophical questions, failing to recognise the significance of the more fundamental psychobiological processes at the root of these experiences. A vast and growing body of knowledge concerning intersubjective attachment and loss, and the shaping of animals' (human inclusive) experiential realities more broadly, elucidates the role of the implicit forces at work for the emergence of grief and spirituality. Simultaneously it evidences that a human-comparable interpretative dimension is not necessary to experience both phenomena deeply.

---

4   Damasio 2006 [1994].
5   Snyder, Bossomaier and Mitchell 2004; Vallortigara et al. 2008.

I did not know any of this when I was standing in Jay's office several years ago hoping she would agree to supervise my dissertation. The dissertation, undertaken part-time, was going to be about *animals*, not grief just yet – grief happened later. The initial plan was rather vague; it was centred on attitudes towards and (mis)representations of nonhuman animals in monotheistic traditions. In the process of gathering research material over the following months, I discovered that someone had already dealt with the question and they did it well. My zeal for attempting a new approach to the same question was minimal so I decided to drop the topic. Then the idea of grief emerged, and both Jay and I were satisfied with the new choice. From the very beginning, however, the point was not the degree itself. The intention to complete the degree was there – it is good practice to finish what is started if possible – but it was not an imperative, at least not for me. In fact, the reason I returned to studies, close to ten years after having moved countries for marital purposes and relinquished my plans to pursue an academic career in historic linguistics and onomastics, was to help me acquire more systematised knowledge of nonhuman animals and the entire cross-species rights and liberation issue. Working on a specific project would give me both the necessary focus, so I did not derail too much, and the opportunity to expand in various directions as we moved along the path into (and hopefully out of) the abyss of cross-species being.

By that stage, I had been calling myself vegan for several years. I read an article in which the author was telling the reader – in that case, me – that I was personally responsible for the suffering nonhuman animals endure and that it was in my power to change that. The whole issue of animal suffering at the hands of humans – an issue that I had considered before, and as a consequence of which I was, at various times, flexitarian, pescatarian and vegetarian, but it had always felt too vast and overwhelming – suddenly appeared 'easy', manageable, something I could do something about right then and there. I turned vegan – quite literally overnight. I considered it a necessary first step, and the easiest thing I have ever done – an 'unexpected pleasure – relief – in the thought that just by *not* doing something we were saving lives,' as my partner phrased it later.[6] Naturally, I wanted to reach out to

---

6   Brooks 2019, 10.

others, share my 'discovery'. However, with time I started to feel that my approach, my advocacy, was inefficient, inadequate, somehow lacking, and that I needed a deeper understanding of questions and processes involved, hence the idea of a research project. Simply telling humans that nonhuman animals suffer and that an important contribution to reduce (if not necessarily eliminate) the suffering was the shift to plant-eating was not a successful strategy. Many sympathised but very few changed their practices. 'That's because we are herd animals,' someone told me recently, citing a blogger they were reading:[7] you cannot take people out of a herd without giving them a new herd – one they can trust and that makes them feel safe and cosy; being alone is scary.

Fortunately, the pro-animal herd is growing rapidly, both in the wider world and in scholarly circles. What only ten years ago was still considered an oddity is now a new normal. For instance, back then the term 'vegan' often had to be explained in restaurants, and when I announced my new research topic at an informal dinner gathering a philosopher inquired, genuinely concerned, how I was planning to deal with anthropomorphism. I did not have an answer at the time; I was hardly aware of the 'problem'. Importantly perhaps, as I realised halfway through my research, at the time I was also only half-aware of the biological, psychological and social comparability of humans and other animals. The aspect of physical pain was clear, the rest less so. Undoubtedly, this contributed to my feelings of inadequacy in advocating for cross-species *equality*.

The beginning of my research project roughly coincided with us moving to a small property on the edge of town. Rescued sheep soon joined our family nucleus of two humans and a dog, and slowly we began to discover that the area was populated with other residents who had been – cautious and hidden – observing and trying to establish the nature and degree of danger the new human-cum-dog settlers may represent. Living with some nonhuman animals and getting to know others, particularly those belonging to substantially unprivileged species, through fostering or assisting in rescue/sanctuary settings, help to imagine a world in which the human is not only removed as the centrepiece of this world but oftentimes becomes redundant. Charlie

---

7   https://jamesclear.com/.

the dog and Henry the sheep kept reminding me of this. Watching them play together was an evocative experience that can hardly be adequately translated into human verbal parlance. To be able to play in the first place they had to negotiate species differences that are not an issue when playing with a member of the same species. They managed that well; they learnt to read intentions and expectations. This involved understanding the significance of various moves within the play context as these same moves change meaning with the change of context. Henry, for instance, had to learn that Charlie's growling was not a threat but an invitation to play; similarly Charlie had to interpret Henry's head-butting move as a friendly gesture rather than a warning signal. Body language aside, they may also have relied on chemical language, which is inaccessible to me as a human with an underdeveloped sense of smell.

It is easy to get carried away by the richness of nonhuman existence. Many other humans must feel the same. It reflects in their writing and other projects. Over time I discovered a vast repertoire of inspiring voices that have, singularly and cumulatively, been helping challenge established perceptions of nonhuman animals and human exclusivism. The new field of animal studies is growing and spreading to various areas of inquiry and creativity. When recently, a senior-year high-school student unexpectedly contacted me wanting advice concerning what studies to pursue to be able to best help animals, I hesitated at first, feeling out of my depth at the prospect of influencing the direction of the life of a complete stranger. After some thoughtful consideration, however, I realised, with a substantial amount of pleasure, that nowadays it does not really matter whether one chooses biology or architecture. Whatever area we work in there will always be opportunity to support the multispecies justice project, and choosing an area we are genuinely passionate about may increase our capacity to help and the quality of the help we can offer.

Far from ideal, the current situation nevertheless feels promising, and of course there are those who came before, who wrote and thought about these questions, paving the way. Mary Midgley was one of them. She published her first book the year I was born, but I discovered her quite late in my research process. How different would my life have been, I wonder, had I grown up reading those books as opposed to others that were perhaps subtly but nevertheless consistently reinforcing the human–nonhuman divide. What would the world be like now had we all been prevalently exposed to works that were more inclusive, less preoccupied with what makes us *special* and more with what makes us open and kind? Even in that case, I suspect, there would be plenty of room left for improvement.

Midgley warned about the futility of the widespread search for human excellences and exclusive properties already in her 1978 book, *Beast and Man*. Her commonsense proposal instead was that what makes humans unique as a species is a conglomerate of attributes and 'the shape of the whole cluster'.[8] The same is valid for any other species, and sharing characteristics with other species does not make one species less unique. Yet the quest continues, and when a phenomenon that was previously thought to be exclusively human is found in other species, humans tend to dig further in an attempt to find differences and human uniqueness in the manifestation of that phenomenon. This is not per se problematic since manifestations can be species-specific, but it becomes untenable when manifestations in other species are not put under equal scrutiny. This may consolidate the prejudice against other animals, which is based on the assumption that humans are indeed a very special animal while all others are not. In the worst but not unusual scenario, differences are simply taken for granted without properly defining the parameters of comparison. The latter often applies in discussions of grief and spirituality.

Much could be said about how human assumptions on grief and spirituality as well as on the connection between the two have led to denying that nonhuman animals possess the capacity to experience either. The idea, in monotheistic traditions, of nonhuman animals not having a soul and a spiritual dimension sets them further apart from humans. This theological proposition continues to influence thought even in contemporary secular circles, in which many humans wish to distance themselves from more traditional forms of religion but do not want to (or perhaps cannot) relinquish spirituality. In this context spirituality becomes defined as a cluster of meanings and understandings of the world and life in it, which may then inform

---

8   Midgley 2002 [1978], 198.

conduct. The contents – cosmologies, ethics, behaviour, etc. – may differ from theistic religiosity, but principles and functions remain comparable. Most humans do not consider nonhuman animals' cognitive capacities sophisticated enough to frame the complexity of the external cosmos and their own internal reality into more or less integrated conceptual systems recognised for human societies, and to appreciate the value, the *sacredness* of it. As a consequence nonhuman animals are rarely considered for their spiritual capacities. There are two principal problems with this position.

First, as Midgley aptly remarked in relation to elephants, 'we can do justice to the miracle of the trunk without pretending that nobody else has a nose.'[9] The fact that nonhuman animals (and, until recently, also humans) have not produced abundant records reflecting intense preoccupation with the construction of such conceptual systems does not mean that their minds are not engaged in meaning-making. Successful navigation through their worlds is to a great extent dependent on their capacity to make some sense of those worlds: to understand phenomena, find connections, predict behaviour, embrace pleasure and cope with adversity, and similar skills (some learnt, some inherited) that absorb the minds of human and nonhuman animals alike.

Second, human conceptual systems or worldviews should not be equated with spirituality. Some of these systems may confer a central role to the experience of a spiritual dimension, and active spirituality may thus be encouraged. However, the system itself remains a product of cognitive appraisal and various extents of cognitive closure (the resolution of a state of uncertainty), whereas spirituality manifests as a propensity of the intrinsically relational non-reflective, experiential consciousness. While capacities for conceptual thinking and consequently cognitive elaboration may differ in degree among animal species – this conceptual muscle being particularly well developed in humans – both human and nonhuman animals have equal access to experiential consciousness, which is fundamental for spirituality. This level of consciousness, as convergent scientific evidence uncovers, is already felt – experientially meaningful – without having to be converted into meaning by some interpretative cognitive functions.

---

9   Midgley 2002 [1978], 198.

However, in discussions of spirituality, human commentators tend to obfuscate these two processes, suggesting (sometimes implicitly, other times explicitly) that the body-based spiritual experience can only acquire meaning through cognitive post-processing and contextualisation within a broader picture of an individual's life. As a consequence, and given that the conceptual world of other animals remains largely inaccessible to humans, nonhuman animals' capacity for spirituality is usually dismissed.

In a similar vein, when considering grief in nonhuman animals, the subjective immediacy of the experience of loss, which is central to the emergence of grief, is often underestimated. The theory and observation of attachment and loss in humans and other animals leave little reason to doubt that what is experienced is similar. However, the potency of bereavement when discussing nonhuman animals' experiences becomes obliterated by commentators asking potentially unanswerable philosophical questions concerning, for instance, awareness of one's own mortality. Asking these questions is not per se contestable,[10] but in these discussions scholarly rigour often becomes lost by presupposing a non-existent universally human understanding of death, mortality and related issues.

Considerations of animal (human inclusive) grief and grief expressions become a problematic exercise if at least two criteria are not met. First, we need to have an adequate appreciation of the psycho-biology modulating attachment and loss. Colin M. Parkes, a long-time student of human grief, pointed out that one of the problems with discussing and analysing grief was the lack of an accepted definition, and suggested to focus on separation anxiety as the core component of grief.[11] Separation distress, which can manifest in innumerable declaratively conscious or unconscious ways, appears to

---

10 To various extents it is also understandable since it appears that many humans fail to embrace their condition of impermanence, opting instead for denialism of their own 'creatureliness' and mortality, with consequences for their attitude towards other animal species. See, for example, Marino and Mountain 2015 and Grušovnik 2018.

11 Parkes 2009 [2006].

represent valid common ground for discussing grief in both human cross-cultural and cross-species contexts.

Second, as commentators we need to step out of our own cultural box if we wish to compare human and nonhuman animal grief and expressions thereof, and/or discuss broader death-related philosophical questions. In the global human context, past and present, in fact, expressions of grief and mourning that may not be considered acceptable in the sterile environment of Western mortuaries are not uncommon. A quick glance across human cultures also uncovers the picturesque fabric of meanings and understandings of life and death.

This research began with a focus on grief. While grief remains the central topic, I have incorporated spirituality into this discussion due to the similarities between the two discovered during the research process, and, perhaps more importantly, due to the possibility of their shared origins in our relation with the vitality of (the rest of) space. I discuss this aspect and the liminality of individual/place in more detail in Chapter 4. With regard to the similarities, they manifest on a procedural and discursive level. Procedurally, both grief and spirituality are intrinsically relational and defined by implicit processes that are primarily felt rather than thought. Spirituality, for instance, is often referred to in terms of self-transcendence, whereby the self is intended to describe the human-grade cognitive, reflective self. However, a closer examination in consideration of scientific evidence uncovers that what in fact enables a spiritual experience is access to the implicit, non-reflective experiential self and the individual's propensity for *self-extension*, or *self-expansion*, as it has also been called.[12] This extending of the self – merging with other selves – is enacted on the self–nonself continuum. According to Shihui Han and Georg Northoff,[13] who researched the phenomenon, this is a more veracious way of viewing self/other–relatedness than the alternative, namely a self–nonself dichotomy. Such self-extension materialises also in intersubjective attachment: when we lose a beloved individual we are effectively losing part of ourselves.

---

12   Nelson 2009, 58.
13   Han and Northoff 2009.

In terms of discourse, as already indicated, both grief and spirituality have either been dismissed as propensities of nonhuman animals or considered weaker in their experiential potency compared to humans' experiences. In both cases, evaluations fail to attend to the subject matter in question, opting instead for considerations of satellite reactions based on the human interpretative domain. The latter is potentially incomparable with other species' interpretative solutions and does not appear critical for the two topics under examination. It is important to notice, however, that many of these 'evaluations' of other species' grief and spiritual experiences and their comparability to humans' experiences tend to be made in passing and to be based on assumptions, rather than being a result of detailed analysis.

An integrated consideration of the above issues, which the present work intends to achieve, is of ethical significance because of the treatment of nonhuman animals whose suffering due to anthropogenic violence (direct or in terms of habitat destruction) is increasing. It also significantly extends the radical interdisciplinary examination of the parameters of the 'human' and the 'nonhuman' that emerging understandings of human/nonhuman animal relations demand.

In my first chapter I outline the path that research into nonhuman animals' lives and subjectivity has taken from the establishment of ethology early in the previous century to current perspectives and approaches, which have changed significantly, particularly in the past twenty years. I review selected material that evidences how ideology on the one hand and proximity on the other have influenced views of other animal species and informed research approaches and focus. This background information is provided to help the reader understand past and current attitudes to nonhuman animal subjectivity in Western discourse, including grief and spirituality, and to contextualise the research of the present volume.

My second chapter looks at intersubjective attachment as the basis for feelings of loss and grief. Physical pain aside, the disruption and/or prevention of intersubjective attachment relations and societal quality more broadly is arguably the major welfare concern with regard to humans' intervention in and manipulation of other animals' lives and relations. This chapter examines more closely attachment theory and the neuro-bio-psychological bases for attachment relations in animals.

The importance and inevitability of attachment relations in animals predicates the experience of grief. Separation anxiety/distress is well documented for human and nonhuman animals alike.

It may be that, reading through Chapter 2, the philosophically savvy readers will find themselves with a vision of French philosopher Maurice Merleau-Ponty putting down his pipe and looking out through the window at the trees of the boulevard. 'Don't you find it a little too mechanistic? Bordering on realism?' he may say, turning towards American lover of wisdom Cynthia Willett, who got momentarily distracted wondering whether the current anti-smoking climate pervading the superorganism is going to reach the old philosopher's perception. 'Yes,' she would probably agree, 'there is the risk that the reader may miss the relationality between individuals implicit in the discussion of psychobiological regulation whereas the concept of affect attunement within attachment dyads bears a stronger suggestion of meaningfulness, emphasising in itself the proto-conversation between animals.' Willett has herself written on grief and spirituality through theories of affect attunement – the sharing but also challenging and altering of moods, affects and desires[14] – which amplifies and complements dimensions of regulation. The absolutely fascinating nature of the material aside, my choice to prioritise the biological imperative of attachment relations is based on my feeling that it leaves little space for questioning the importance of the presence and quality of these relations across relevant animal species. I also trust that the reader will not fail to sense the richness of social exchanges behind the organismic processes discussed.

The psychologically informed reader, on the other hand, may wonder whether the length of Chapter 2 and detailed presentation are justified given that attachment research has received substantial public recognition, leading to growing awareness of the various aspects and processes. Over and over again, however, I have found people – non-psychologists – marvel at the available information, especially in relation to nonhuman animals, which helps them towards a new understanding and acceptance that the experience of grief truly is

---

14   Willett 2014, 88–9.

comparable between human and nonhuman animals, just as it helped me when I first delved into this research.

Most considerations of animal grief presuppose a non-existing panhuman homogeneity in key grief- and death-related philosophical questions and in the expression of grief. A closer examination of both issues in the human realm challenges these assumptions. In its first part, my third chapter examines the question of understanding death from a biological (physical non-returnability) and – to a limited extent – cultural perspective (different conceptualisations of death), and what this may signify for nonhuman animals and grief. The second part of Chapter 3 explores a variety of grief expressions as well as repressions in human societies (from delayed personhood to cannibalism), challenging the widespread misconception that if nonhuman animals do not exhibit mourning that is comparable and recognisable to the (Western) human, they do not experience grief.

Turning from a focus on grief to that of spirituality, my fourth chapter explores spirituality as a manifestation of the implicit, experiential, non-reflective self. From this perspective, I suggest that spirituality is ontologically distinct from religion and religiosity; the latter tend to be characterised by a strong cognitive closure component, and may to various extents help individuals cope with grief. Nonhuman animals also engage in cognitive categorising and closure; nevertheless, they likely do not have an elaborated cosmology like many humans do, which may mean that in the absence of this coping strategy grief may be harder for them. Spirituality, on the other hand, is experienced by human and nonhuman animals alike, and its developmental origins could be traced in place attachment, which has been suggested to have been the evolutionary precedent of intersubjective attachment.

Returning to the topic of grief, the final chapter discusses vicarious and disenfranchised grief that some humans feel in relation to nonhuman animals, and the various forms of bearing witness to their suffering, including the emergence, in recent years, of organised public vigils. As they stand in silence, commemorating the lives and deaths of animals whom the human society perceives as mere commodities, the mourners bear witness to a different reality: one of profound suffering of the nonhuman victims themselves, the humans who refuse to look the other way and those who are not given the choice (for example, slaughterhouse workers) – a reality of sights, sounds and smells that have over time disappeared from public perception and hence awareness, and that we now have the opportunity to revisit and re-organise.

There is something freeing about perceiving oneself as a fragment of the whole. You are not insignificant even though you are very small and nothing special, embedded in living like the rest of life, equal to other animals and the rest of existence. You are here now, like they are, no-one's fault, no-one's merit. We just are – individuals of all kinds – freckles in the continuation of existence, specific (and mostly accidental) manifestations of being.

In the following pages I recount what I have learnt over the years, through theory and practice, about ourselves and other animals. This new knowledge and understanding have had an immense impact upon my life as I went from being a human who needs to protect 'animals' (as less privileged forms of sentient existence), to being an animal among many who needs to learn new ways of living and relating, within myself and towards the outer world.

1
Animal subjectivity

We need another and a wiser and perhaps a more mystical concept of animals. Remote from universal nature and living by complicated artifice, man in civilization surveys the creature through the glass of his knowledge and sees thereby a feather magnified and the whole image in distortion. We patronize them for their incompleteness, for their tragic fate for having taken form so far below ourselves. And therein do we err. For the animal shall not be measured by man. In a world older and more complete than ours, they move finished and complete, gifted with the extension of the senses we have lost or never attained, living by voices we shall never hear. They are not brethren, they are not underlings: they are other nations, caught with ourselves in the net of life and time, fellow prisoners of the splendour and travail of the earth.
    — Henry Beston, *The Outermost House*, 1928[1]

Throughout his life, Charles Darwin suffered from a series of symptoms indicative of hyperventilation syndrome. This condition can have a psychological basis, being a result of stress, trauma or grief. Given that Darwin lost his mother when he was only eight years old and that his father forbade the children to speak about her, John Bowlby,

---

1   Beston 1949 [1928], 25.

15

the pioneer of attachment theory as well as Darwin's biographer,[2] concluded that Darwin's condition derived from his unresolved, or complicated, grief for his mother.[3]

Complicated grief is discussed in more detail later in the book. Suffice to note, at this point, that forms of complicated grief may develop if a more normative course of mourning is somehow prevented, as has been suggested happened in Darwin's case. It may also develop as a consequence of insecure attachment styles. The latter likely occurred in the case of Chickweed, who was brought to the Eastern Shore Sanctuary in Maryland (now VINE Sanctuary in Vermont) along with his sister Violet when they were tiny chicks. Following the death of Violet, whom he was very bonded to, Chickweed 'for the next several weeks, would return to stand silently at the place from which he had last seen her,' explains the sanctuary manager;[4] 'he became angry and would rage around the yard every day,' and at night remained alone in the coop 'drooping with sadness'. Over time his anger receded but he never truly recovered. This comparison between Darwin and a chick may raise some eyebrows, though perhaps least of all those of Darwin himself had he heard about Chickweed's grief, since the nineteenth-century scientist showed little doubt that nonhuman animals experience rich emotional lives.[5] Biology and evolutionary continuity aside, there is another connecting element between Darwin and chicks: both, in their own distinct ways, significantly influenced the emergence and formulation of Bowlby's attachment theory, which is fundamental for discussing and understanding loss and grief across animal species.

The evolutionary perspective coupled with improved research technology has, over the decades since the theory was first proposed, enabled increased insight into processes underlying attachment and loss in animals. These findings have been continuously used in attempts to clarify human conditions and provide help for humans when necessary. *Anthropodenial*,[6] or blindness to the humanlike characteristics of other

---

2   Bowlby 1990.
3   Reported in Fraley and Shaver 2016, 40.
4   Reported in Hatkoff 2009, 31.
5   Darwin 1872.
6   de Waal 1997.

animals, on the other hand, has, in Western science and philosophy, consistently precluded embracing the possibility of the existence of human-comparable subjective dimensions in other animals, including the feeling of grief. This conceptual barrier between humans and other animals is, however, progressively coming apart.

Historical perspectives and cognitive biases

Double looking-glass

The 'cross-race' or 'own-race' effect denotes the increased ease with which humans recognise faces from their own race compared to those from other races.[7] This phenomenon, whose selective aspect is plastic and may thus disappear following sufficient exposure to other races,[8] is equally applicable to other species as well as interspecies contexts. For instance, the phenomenon played an interesting trick on scientists who were trying to establish chimpanzees' capacities for face recognition. Since human faces are 'clearly' (that is, according to the scientists) more distinct than chimpanzee faces, the scientists decided to use human faces to test chimps' face recognition. The chimps turned out to be terrible at it, which of course induced concerned scientists to conclude that this particular capacity is not very well developed in chimps. It took a little while and some out-of-the-box thinking for someone, namely Lisa Parr from the Yerkes National Primate Research Center in Atlanta, to test the chimpanzees again, but this time using chimpanzee instead of human faces. Lo and behold, primatologist Frans de Waal reports, the chimps excelled at it.[9]

Generally speaking, nuances in appearance, facial definition and expression, body language, and even, or particularly, intracommunity interactions of other animal species can easily be missed if humans are deprived (or deprive themselves) of participation in the animals' worlds – be it direct participation involving interaction, or participation by

---

7 Bothwell, Brigham and Maplass 1989.
8 Sangrigoli et al. 2005.
9 de Waal 2016, 18–9.

observation alone with the attempt at attunement, albeit at a distance, with the animals and their lives.

In some human cultures, such as our Western culture, humans' ideologically informed species segregation in their choice of corporeal comestibles leaves certain animals particularly vulnerable to depersonalisation and devaluation of their individual and social features and competencies. This is reflected in the lack of attentional focus on these species in scientific inquiries as well as in the attitude of the general public towards these species, both of which determine political (in)action. For example, an alarmingly high number of humans ask whether the rescued sheep I live with exhibit distinct personalities. Human people living with individuals belonging to other profoundly instrumentalised and depersonalised species, such as hens, pigs or cows, address similar inquiries. Viewed as 'quasi-artefacts', to borrow Freya Matthews's unflattering but widely condoned description, who have lost their sovereignty, 'farm' animals 'owe their existence to us [...] We are obliged to care for them but we also have certain rights over their destiny.'[10] While domestication was wrong when it occurred and it would be equally wrong to try to domesticate wild animals in the future, this does not mean, Matthews argues further, that the farming of species that are already domesticated is wrong today; the 'pact' that many humans believe we have with these species permits us to use them for our own purposes. However, this idyllic and idealistic view does not reflect the reality of animal agriculture as a system nor that of individual animals trapped in the system and victims of what I call ontological genocide[11] – a genocide based on the appropriation of a psycho-socio-biologically complex life and its conversion into a mere commodity.

It is not unusual – quite the opposite, in fact – for human societies to assign culturally edible bodies among the physically edible bodies in their socio-natural environments. For example, traditionally, for the Amazonian Wari' people the physically edible bodies of deceased in-laws represented culturally edible bodies. The entire socio-emotional spectrum of action and reaction at the occurrence of death, including

---

10   Mathews 2013, 264.
11   Brooks Pribac 2016.

coping with grief, informed by this ritual was discontinued following Western intervention and the introduction of burial practices, which some Wari' people still find discomforting.[12] Animals in some cultures, such as the Indigenous Australian totemic cultures, may be excluded from groups of animals considered as culturally acceptable corporeal comestibles. In most Western cultures pigs, cows, chickens and some other species qualify as food, but others, such as dogs and cats, are spared because of their assigned privilege as companion animals. Stemming from this culturally shaped normativity, Westerners are quick to condemn as 'savages' those Asian communities that consume dogs and cats (and of course those humans practising cannibalism), ignorant of (or perhaps simply ignoring) the fact that the Western species segregation is equally arbitrary and a result of the ideology of *carnism*, as Melanie Joy terms it.[13] Joy describes carnism as an invisible belief system that enables us to see both flesh consumption generally and consumption of the flesh of certain animals specifically as something 'natural'. 'We don't see meat eating as we do vegetarianism – as a choice, based on a set of assumptions about animals, our world, and ourselves,' Joy writes, continuing: 'We eat animals without thinking about what we are doing and why because the belief system that underlies this behaviour is invisible.'[14] The so-called mere exposure effect enables us to internalise and consolidate the notion that some animals are here for us to eat them, while principles of the system justification theory help maintain the status quo.

The mere exposure effect describes a psychological phenomenon, vastly used and abused in advertising, by which the exposure to stimuli determines one's drives. The more one is exposed to specific stimuli and therefore becomes familiar with them, the more one likes them and becomes attached to them, particularly if exposure is accompanied by positive experiences or at least by the absence of negative ones.[15] Exposure-determined attraction has been documented as valid for all the animals who have been studied, and ranges from food to places and

---

12 Reported in Harvey 2005, 157–60.
13 Joy 2010.
14 Joy 2010, 29.
15 Panksepp 1998, 259.

even to ideas. It is not uncommon for humans exposed to certain ideas to begin to prefer those ideas over unfamiliar ones proposed by others, and to often eagerly contradict the latter prior to seriously considering them.[16] A study by Gallate and colleagues addresses the statistically unwarranted association of Arabs and terrorists in mainstream Western psyches, media-fuelled and exacerbated after 9/11.[17] Researchers managed to reduce the prejudicial score of the tested (human) subjects via non-invasive brain stimulation. While the neural basis for prejudice remains relatively unknown, prejudice seems to sit comfortably within the context of the highly automatic categorisation processes in the brain,[18] to which we return below. The latter can be beneficial in simplifying the complexities of the world, yet quite detrimental when oversimplification and overgeneralisation occur with prejudicial segregation as a result.[19]

Reasoning processes are not exempt from the influence of the environment, including the mere exposure effect: indeed, quite the opposite. Mere exposure contributes to the shaping of the conceptual and emotional framework, determining to a certain extent what one sees, how one sees it, and how one acts in relation to themselves, conspecifics (organisms belonging to the same species), other species and the rest of 'nature'.

The system justification theory,[20] on the other hand, predicts humans' tendency to perceive the larger system one is embedded into, and dependent upon, in a positive light regardless of how harmful the system may be. The research in this area focuses primarily on the human intra-species context and the puzzling desire to keep the status quo, even by groups and individuals who would obviously directly benefit from change. Nevertheless, system justification can easily be observed in relation to the treatment of nonhuman animals, with the wider public believing (or wanting to believe) that the system has provisions in place to ensure that nonhuman animals do not suffer

---

16   Panksepp 1998, 259.
17   Gallate et al. 2011. See also Merskin 2004.
18   Gallate et al. 2011.
19   Gallate et al. 2011.
20   Jost and Banaji 1994. For a review see Jost, Banaji and Nosek 2004.

on their way 'from pasture to plate'. Undercover investigations as well as testimonies from former meatworkers (we return to them later in the book), however, evidence the profound and relentless suffering of these animals. Further, the code of accepted farming practices, which is supposed to ensure some basic welfare standards for 'farm' animals, provides guidelines but these are recommendations only; they are not statutory, leaving 'food' animals virtually without protection.[21]

The position that domestication occurred in the distant past and that at present we are simply left with ancestral 'artefacts' which we have an obligation to look after is both inaccurate and highly misleading. Animals trapped in the exploitative systems sprung with and promoted by domestication are subjected to ongoing manipulation to increase production and profit. Over the past sixty years the size of a broiler chicken has quadrupled,[22] and the bodies of other animals, enslaved for human dietary choices, such as turkeys and pigs, have also substantially increased in size, leaving their underdeveloped bones and internal organs struggling to keep up with the unnatural body mass they are supposed to support but often cannot. 'They are bred to be slaughtered at six months of age,' explained a sanctuary owner whose hallway has been taken up by a disabled rescued pig suffering from apophysiolysis,[23] adding: 'This and similar conditions are not uncommon, particularly in breeding sows, who are obviously kept alive past the six months.'[24]

Sheep are another species under constant attack. In Australia, about one in four lambs (fifteen million annually) dies from exposure. To address production loss, the industry is engaging in genetic manipulation aimed at increasing the number of lambs per birth, pushing the mothers well beyond their physical (and undoubtedly emotional) limits, and likely increasing death rates of lambs, although overall the number of lambs to be turned into meat may increase.[25]

---

21 Mark 2014.
22 Zuidhof et al. 2014.
23 Fracture of the ischial tuberosity of the tail bone.
24 Ksenija Vesenjak, personal conversation, July 2015; Vizcaíno et al. 2012.
25 Animals Australia 2015.

The second point, which is also often ignored by proponents of animal agriculture ('humane' or other), is the recognition of animals' interpersonal bonds and social structures. Animals who are allowed a measure of physical and psychological freedom exhibit social preferences[26] and can form deep, lasting bonds,[27] inclusive of, but not limited to, the parent–child dyad. When humans claim rights over animals' destinies and enact such rights, for example by choosing to kill someone from the community, they may be breaking up meaningful interpersonal relations as well as affecting the animals on a societal level.

The third point to consider is Donald Broom and Ken Johnson's assertion of the perpetual victory of evolution and adaptation over domestication.[28] They famously wrote that millions of years of evolution and adaptation cannot be overridden by a few thousand years of domestication and a few decades of close confinement. They cite, specifically, the unlikelihood of a hen adapting to live in a cage any time soon regardless of the level of genetic manipulation involved, and the Australian wild boar – communities of domesticated pigs who have strayed from farms and successfully returned to a wild state. More recently, communities of rescued hens and roosters at the VINE sanctuary in the United States achieved something similar, and now they inhabit the nearby forest, living a 'wild' life, free from human intervention.[29]

In her book *Interspecies Ethics*, Cynthia Willett reminds the reader of African-American abolitionist Frederick Douglass's doubts that an appeal to the moral sentiments of white people would suffice to abolish black slavery:

> White people could not generate sympathy for a slave unless that slave asserted some significant degree of agency and demanded, through that assertion of agency, recognition from others [...] A display of vulnerability and an appeal for sympathy do not suffice to generate the solidarity that an egalitarian political ethics requires.[30]

---

26 Bode, Wood and Franks 2011.
27 e.g. Holland 2011.
28 Broom and Johnson 1993, 33.
29 jones 2010.
30 Willett 2014, 38.

Appeals to sentiments do not appear to work even between human groups, and are thus unlikely to lead to liberation of nonhuman animals. Yet nonhuman animals keep reminding us of their desire for self-determination. Could re-wilding, shown by the aforementioned pigs and hens, be the strongest demonstration of agency that nonhuman animals could offer, along the lines of Douglass's thinking, to convince humans of their equality and longing for freedom – a kind of peaceful revolution, Mandela-style rather than Orwellian? Aren't the animals who have freed themselves from human ownership demonstrating just that: the desire to be free and the capacity to live this freedom? This includes negotiations with a highly complex socio-natural environment that go beyond food acquisition and reproduction success, requiring high levels of cognitive and emotional sophistication, which is increasingly being recognised in nonhuman animal societies. (Admittedly, many domesticated animals would not be able to survive in the wild, mainly due to the genetic mutilations humans have subjected their bodies to: most domesticated sheep breeds, for instance, need to be shorn regularly, and the males of domesticated turkeys develop such large breasts that they can no longer copulate naturally.[31])

The answer, however, appears to be negative. Free-living nonhuman animals ('wildlife') do not fare much better compared to captive animals, and there is certainly little admiration for Australian re-wilded pigs, who are considered a pest and hunted in ways that cause profound suffering.[32] Free-living nonhuman animals are also subject to a 'pact' they have never agreed upon, and the value of their lives and communities is also to a large extent filtered through the lens of various human needs and expectations. From pest to emblem, the celebration of their 'wildness', when it occurs, remains largely symbolic with little appreciation for their individual and societal wellbeing and little political will to promote it.

The planet is currently in the middle of what scientists hold to be the sixth mass extinction, the first mass extinction caused almost

31   https://bit.ly/3l3sirs
32   So-called 'pig dogging' describes the practice of releasing packs of trained dogs to track and maul pigs in the wild.

entirely by human activity, as the Center for Biological Diversity notes on their website.[33] Besides the often-cited human voyeuristic interest in species preservation, the primary tragedy of extinction is the many individual animals' suffering before their own and their conspecifics' lives are permanently lost. Moreover, extinction does not exclusively affect the species in question, but also the entire multispecies ecology (also composed of individuals) that the vanishing species is embedded into, including participating human communities, as van Dooren elaborates in relation to vultures in India.[34] Neither does extinction need to be imminently physical to have a significant impact on individuals and communities. Bradshaw reminds us of the psychological extinction of animals in the wild, whereby the outer form remains the same but inside, due to anthropogenic violence, including habitat loss and massive deaths, 'they bear scant resemblance to the past'.[35] This, with its cross-generational transferability, may change their inner landscape and outer relational patterns forever, much as in the case of animals exploited in agribusiness.

The physical pain suffered by nonhuman animals through anthropogenic violence both in exploitative industries and in the wild is immense and, as such, of deep concern. Spreading awareness of this pain is, understandably, central to all animal-protection advocacy, be it abolitionist or regulatory in nature. However, the focus on physical pain can dim other equally important aspects of an animal's being and encourage their further objectification and the notion that enslavement, exploitation and extermination might be better justified should that pain be absent or reduced. The widely practised instrumentalisation of animals (whereby the animal becomes an object, a tool to be used for human purposes) with its constant and systemic attempts to silence them represses the expression of their being – a being that the human has appropriated and expects to function principally for human interest. It also largely[36] ignores aspects of

---

33   https://bit.ly/3naKIYS
34   van Dooren 2010.
35   Bradshaw 2014, n.p.
36   Increased sensitivity to nonhuman animals as complex socio-biological
      entities has sprung a 'movement' within animal-welfare circles advocating for

positive sentience – that is, rewards and pleasures [37]– and the need for agency, for self-determination, which are equally important for wellbeing. The recent move towards a more integrated understanding of nonhuman animals' individuality and relationality shows that the extent of suffering is much greater than previously believed, but coupled with the appreciation of the cognitive bias that has brought us to the present state of affairs, it may also inspire change.

---

greater consideration of positive sentience by factoring in environment- and sociality-focused needs and potentialities as markers of wellbeing along with the absence of negative stimuli (see, for example, Mellor 2015; Yeates and Main 2008). While this is certainly a refreshing perspective, its practical limitations are not insignificant. The observance of positive sentience could lead to the emergence of a boutique industry, which could serve its own purpose and possibly benefit a small number of animals, but it appears to be an unrealistic option to solve the current pressing issue of animal wellbeing. Such industry would be unable to meet the current demands for animal products by the ever-growing human population for various reasons, including our planet's space limitations, and it could certainly not match the current financial affordability of animal products, which in an environment of growing economic pressure is not a negligible factor. Further, apart from animals forming the heart of the production line to whom regulations concerning positive welfare could apply to various extents, the current establishments also comprise animals that are deemed completely superfluous by the businesses in question (e.g. male chickens in the egg industry); as such, these animals need to be disposed of in a timely fashion and in ways that are financially the least impactful. Higher 'humane' standards would have to address the issue of these superfluous animals as well as the question of slaughter of both animals raised for meat and so-called spent animals in other exploitative sectors of agribusiness. Ultimately, animal agriculture is a business and its existence depends on its profitability. If the latter is challenged by welfare standards, the business will either cease to exist or it will find ways of disguising or underplaying the abuse. This is already happening with issues concerning basic physical painism and is also reflected in the widespread misleading advertising of so-called free-range and similar settings implying (but not necessarily implementing) higher welfare standards to please the public and its growing awareness of animal use and abuse.

37 Balcombe 2009.

On sheep and other primates

Groundless human projections continue to be evoked to dismiss proposed nonhuman animals' characteristics capable of disturbing the biblical foundations[38] promoting human supremacy, upon which the Western mind was built and within the framework of which it continues to operate, even in secular circles. But, increasingly, students of animals' intra- and interpersonal competencies agree that accusations of such projections ('anthropomorphising') are often premature and uninformed; they emphasise instead the importance of participating in the animals' world, understanding how it works, otherwise, as fish expert Culum Brown points out, 'chances are, you either ignore them entirely or you misunderstand them'.[39] Further, the greater the human economic and ideological investment in the instrumentalisation of particular species, the less is the motivation to explore and understand the life and being of these species, their social fabric and individuality within it.

In relation to social intelligence, primatologist-turned-sheepologist Thelma Rowell draws parallels between methodological faults (and consequent misleading results[40]) of early primatology – before

---

38   The Bible is a complex text open to interpretation. Most interpretations that have informed Christian and Jewish morals support instrumentalisation of nonhuman animals and their use for humans' need and greed. Genesis 1:26 is often cited to justify such use: 'And God said, Let us make man in our image, after our likeness: and let them have dominion over the fishes of the sea, and over the fowl of the air, and over the cattle, and over all the earth, and over every creeping thing that creepeth upon the earth' (King James version). For a discussion and alternative interpretations see, as an example, Linzey and Cohn-Sherbok 1997.
39   Brown 2014, n.p.
40   An interesting example that sparked and perpetuated the myth of aggression and competition as a societal norm in primates' communities comes from research on baboons, albeit the erroneous research conclusions were less a result of inadequate observation than they were of the non-normative and traumatogenic captive environment the research was conducted in. As Despret summarises, from the observations of baboons in the London Zoo in the late 1920s by zoologist Solly Zuckerman, a thesis was developed that presupposed dominance-hierarchy as the main principle of social organisation in primate societies generally. This thesis prevailed for several

primatology shifted from ethological to more anthropological research methods[41] – and the continuing ethological approach in research of sheep communities. In the words of philosopher Vinciane Despret:

[A]s far as their social expertise is concerned, these animals are certainly on a par with apes. To put it simply, they are organized – so much so, in fact, that they warrant the title recently awarded to dolphins, hyenas and elephants, of 'honorary primate' […] Of all animals, sheep are precisely those that until now have been given the fewest chances. They have been the victims of what Thelma Rowell calls 'a hierarchical scandal' in ethology: 'we have given primates multiple chances; we know just about nothing about the others.' Of course we know things about them, but clearly those things are incomparable to what we know about apes. The more research advances, the more interesting the questions about apes become, and the more these animals turn out to be endowed with elaborate social and cognitive competencies. By contrast, questions about the others still primarily concern what they eat.[42]

A similar criticism of 'primate chauvinism'[43] was recently advanced in relation to the social and cognitive intelligence of fishes, citing largely Redouan Bshary's work.[44] In essence, humans have blatantly taken everything away from nonhuman animals – to resort to a hyperbole – and turned them into unfeeling, unthinking objects for humans' own convenience. Now nonhuman animals have to depend on human research ingenuity and attentional focus to prove humans wrong, and possibly change their attitudes towards other animals. Fortunately, this has begun to happen on various levels: from the increase in number of sanctuaries and micro-sanctuaries that provide forever homes to

decades to the extent that when the dominance principle could not be observed in a particular primate community, the apparent absence of it would be conceptualised as 'latent dominance' (Despret 2009).

41 i.e. long-term research focusing on social rather than food questions (Despret 2005).

42 Despret 2005, 360–1.

43 de Waal, quoted in Abbott 2015, 414.

44 Abbott 2015.

farm animals and opportunities for exploration and appreciation of various aspects of their being, to more theoretical approaches, such as Lori Marino's *The Someone Project*, which investigates the scientific literature on the minds and being of some of the most neglected and repressed species.[45]

Nevertheless, such a 'depleted' version of nonhuman animals was the baseline when more systematic research of nonhuman animals began in the early twentieth century with the focus on animal behaviour. Behaviourism, as the practice of conveying exclusively what one could see, ecologist Carl Safina notes, developed as a necessity to establish the study of animal behaviour as a science at a time when brain science was in its infancy and little systemic observation had been made of free-living animals conducting their normative lives.[46] This 'objective' approach was also intended to dispel many myths surrounding nonhuman animals, stemming from centuries of folklore and superstition, portraying animals as caricatures of human vices and virtues (for example, sly foxes, evil snakes, etc.). 'In establishing the study of behaviour as a science,' Safina writes,

> it had originally been helpful to make 'anthropomorphism' a word that raised a red flag. But as lesser intellects followed the Nobel Prize-winning pioneers [Konrad Lorenz, Nikolaas Tinbergen and Karl von Frisch], 'anthropomorphism' became a pirate flag. If the word was hoisted, an attack was imminent.[47]

Behaviourism managed to instigate and consolidate the fear of anthropomorphism, a fear that (even though receding across disciplines) remains widespread today. The objectivity that behaviourism strived for, however, fell short of expectations. There is always more than meets the eye, and the nature of the attention the observer applies also plays its part. This can lead, for example, to ignoring species culturally/ideologically deemed uninteresting, such as

---

45   Marino and Colvin 2015; Marino 2017; Marino and Allen 2017; Marino and Merskin 2019a.
46   Safina 2015, 26.
47   Safina 2015, 27.

sheep, or paying excessive attention to certain behaviours while ignoring others of equal or higher relevance for the overall understanding of an observed individual or community. In her book *Animal Friendships*, zoologist Anne Innis Dagg laments that for a long time research focused on aggressive and reproductive behaviours among nonhuman animals and ignored the less 'exciting' though more regular congenial relations, a focus that has undoubtedly contributed to the heavily distorted picture humans still nurture of other animals' lives and relations.[48] It is telling, in fact, that the word 'animal' is considered one of the worst insults one can offer to a fellow human, and if this is the case, if the worst thing one can be called, or treated like, is an animal, where does it leave nonhuman animals, asks Debra Merskin.[49] Marc Bekoff, who has written extensively about nonhuman animals' moral and emotional lives, concurs that cooperation and convivial behaviours are more common.[50] So does Donald Broom, who also reminds us that when considering sociality and morality within animal communities it is critical to take into account not only what individuals do but also what they do not do. In fact, '[m]ost altruistic behaviour involves refraining from doing things which would be easy to do but which would harm others, even if the perpetrator might benefit in some way from doing this'.[51]

Ultimately, the observed phenomena cannot escape subjective evaluation and interpretation as humans are faced with choices of actions that impact upon nonhuman animals. For example, deciding that nonhuman animals' interpersonal relationships do not matter is no more objective than deciding they do matter. The bond between mother and infant is a well-recognised phenomenon, both in scholarly literature and in popular knowledge. The disruption of this bond can lead to production loss in animals exploited for their bodies and secretions; therefore effort has been put into research attempting to minimise the inconvenience, resulting, for example, in various available techniques of forced weaning.[52] Even when it comes to 'commodities'

---

48   Innis Dagg 2011.
49   Merskin 2018, 18.
50   e.g. Bekoff and Pierce 2009.
51   Broom 2003, 40.
52   e.g. Schichowski, Moors and Gauly 2008.

such as farm animals, the bond is obviously there, but, production loss aside, by and large the bond does not matter.

Most of the twentieth century remained scrupulously 'anthropophobic'. In 1992, commenting on Donald Griffin's book *Animal Minds*,[53] a scholar even published a warning in the journal *Science* to those who wished to study nonhuman animals' consciousness, stating that it 'isn't a project I'd recommend to anyone without tenure'.[54] Nevertheless, not everyone was willing to comply with convention. In fact, Griffin himself and Gordon Burghardt went against the current when, in the 1970s and 1980s, they began advocating for the inclusion of subjective mental experiences of nonhuman animals into the field of cognitive ethology.[55] Their 'digressions' were met with substantial resistance; Griffin's 1984 book titled *Animal Thinking*, for instance, was described by a critic as 'the Satanic Verses of Animal Behavior'.[56]

Over the following decades an increasing number of scientists has embraced the idea of nonhuman animals' subjective lives, enabling animals' subjectivity to grow from a forbidden topic into the widely popularised subject that it is today. Even nonhuman animals' psyches are not a taboo anymore.[57]

More integrative views and healing approaches, which recognise a human-comparable psyche, such as trans-species psychology proposed by G. A. Bradshaw,[58] work on the premise that animals (including humans) are born with specific neurobiological dispositions that require specific socio-natural environmental input for the subject to develop as physically and emotionally balanced. The disruption of the biological and/or historical normative (for example, by anthropogenic interference) affects the delicate balance that has slowly emerged through the species' evolution and the manner in which this predicates optimal developmental and living conditions. This leads to

---

53 Griffin 2001 [1992].
54 Yoerg 1992, 831.
55 Griffin 2006, 482.
56 Reported in Griffin 2001 [1992], x.
57 e.g. Bradshaw 2009; Dasgupta 2015; Ferdowsian 2018.
58 Bradshaw 2005; 2009.

compromised wellbeing and, when the stressors intensify to unmanageable levels, to the emergence of severe psychological scars and trauma. This not only affects the wellbeing of the individual in question but, via trans-generational transfer, also impacts on posterity.[59] The trauma imprinted in the animal's subconscious can only be adequately accessed and reorganised (aiming at healing) by providing a secure environment where the affected animals are given the opportunity to re-construct their unconscious[60] and re-create themselves as new, mentally balanced individuals. This is a space that allows and encourages agency and mutual respect as opposed to imposing control. The traditional behavioural approach attempts to correct behaviour through positive or negative reinforcement (rewards and punishment). However, proponents of the trans-species approach maintain that, just as for humans, focusing on behaviour in other animals and attempting to correct behaviour alone does not address or eliminate the root of the problem. As a consequence the wellbeing of the animal(s) continues to be compromised even though the behaviour may appear more in tune with humans' expectations: broken on the outside (behaviourally), but likely broken inside, too. *Phoenix Zones* is one of the most recent tributes to the cross-species psyche. In this book, Hope Ferdowsian, a medical practitioner who has worked extensively with human victims of violence, draws parallels between humans and other animals' shared psychic vulnerabilities as well as capacities for recovery that can materialise, for instance, in the safety of a sanctuary.[61]

Sanctuaries for rescued nonhuman animals offer valuable insight into the complexity of the nonhuman residents' mind and psyche. The primary purpose of these sanctuaries is not research per se, but an attempt to provide a safe environment free of physical and other suffering as well as encourage positive sentience and self-determination.[62] In order to achieve these goals, however, informal

---

59 e.g. Bagot and Meaney 2010; DeGregorio 2012. See also Clark et al. 2014 for pain sensitisation in sheep.
60 Schore 2011.
61 Ferdowsian 2018.
62 Clearly, there are limitations to what an environment that ultimately remains a captive one can offer. Nevertheless, a move away from the standard 'refuge + advocacy' sanctuary model and towards sanctuary as 'intentional community'

but nonetheless meticulous research and observations are de facto being carried out on a daily basis for the entire duration of the animals' residency at the sanctuary. The relational dynamics characteristic of these settings offer alternative modalities of knowing and understanding animals, which inform methods of care as well as advocacy. Considering the difference between true wellbeing and 'welfare' – a term that has largely grown to denote attempts to reduce animal suffering under exploitative conditions – long-time activist and rescuer Patty Mark suggests that it may be easier to understand the extent of the violence and deprivation animals endure in an exploitative context by considering these same animals in a sanctuary environment after they have been rescued. 'The damage becomes much more evident when the animals are at last allowed autonomy,' she says, 'when they are given the freedom, for example, to *not* be touched,' picked up, restricted, immobilised by a foreign hand, 'when you watch them protecting and nurturing their bodies and selves like we do our own' and 'slowly heal physically and psychologically – sometimes it takes years – beginning to enjoy life and friendships with other animals, including humans'.[63]

The intersubjective space of being together, created by the partners (the rescued animals and their carers), enables a relatively fluid transfer of information (albeit of a nonverbal nature) and relationship-building. Dismantling the prejudicial barrier based on culturally primed species segregation uncovers new foundations that enable a more comprehensive understanding of other animals, promoting empathic recognition and informing humans' attitude towards them. 'All animals,' Tom Regan reminds us, 'are somebody – someone with a life of their own. Behind those eyes is a story, the story of their life in their world as they experience it.'[64] The ability to hear these stories is critical

---

where the residents' participatory role in the life and decision-making of the sanctuary is enhanced (for a discussion see Donaldson and Kymlicka 2015) would secure greater self-determination, general wellbeing as well as facilitate change in humans' perception of these animals.

63   Mark 2014, 107. See also Ferdowsian and Merskin 2012 for a discussion on parallels in sources of human and nonhuman animal physical and psychological suffering.

64   Regan 2004, n.p.

to ensure the best possible care and physical as well as emotional and mental rehabilitation. Aside from strictly medical aspects, such care has to take into consideration a myriad of species-specific properties, both cultural and natural, along with, of course, the personal specificities of the individual in question.

As a consequence of this subjective encounter, of living and being together with other animal species, and in certain circumstances being able to experience their group dynamics, sanctuary caregivers (including humans who provide sanctuary/home to singular companion animals, like cats and dogs) may develop capacities of seeing, hearing and understanding animals in ways that many other humans cannot. Unlike humans who participate in the exploitation of other animals and whose vision is by definition blurred as a consequence of this utilitarian 'relationship', sanctuary caregivers are freer of the mental and practical limitations that such instrumentalisation entails and their interest is focused on the wellbeing of the animals in care. Most of them are also much less constrained by the doctrinal requirements that govern Western science, which is itself embedded in a tradition replete with interpretational and methodological errors, some of which are noted in this chapter.

Mindless anthropomorphising would not just be a futile process, but it can also adversely impact the animal in care. It is essential to create a space which, while recognising species-specific characteristics, allows the development of an adequate level of intercommunication in which the human *listens*, as a non-passive recipient, to individual nonhuman animals (and/or a community of animals) who are *telling* their story. Each animal is a product of nature and nurture, that is, the genes and the socio-natural environment they grew up in and are embedded into; as such, each animal is unique, as is their individual story and needs. The failure to recognise this lies not in the animals' lack of individuality and complexity but in the lack of time humans spend with them, as marine-mammal expert Toni Frohoff indicates in relation to whales,[65] or in approaching them with a predetermined,

---

65 Cited in Siebert 2009, n.p.: 'A distinctive aspect of the new cognitive
    revolution that Toni Frohoff spoke to me about is that scientific facts, of all
    things, are now freeing scientists like herself to be more expansive storytellers

culturally biased, oftentimes voyeuristic attitude that does not allow one to be open to truly *hearing* them. What emerges from this process is an extremely complex picture of animals' psychosocial existence, congruent with various theoretical frameworks discussed throughout this volume, indicating the need for a paradigm shift, a shift unlikely to happen if the attention remains on how to justify the use of animals instead of on the animals themselves.

Ultimately, 'what is a "human" emotion?' asks Safina. 'When someone says you can't attribute human sensations to animals, they forget that human sensations *are* animal sensations. Inherited sensations, using inherited nervous systems,'[66] or as de Waal puts it: 'emotions are like organs' and we share them with other animals.[67] The situation is very similar with the psyche more broadly. The reason we understand and are able to address certain (human) psychological disturbances is decades of invasive studying of nonhuman animals' brains. While we continuously infer this information unidirectionally from nonhuman to human animals, bidirectional inference (from human back to nonhuman animals) remains labelled as 'anthropomorphism' largely on ideological grounds.[68] After all, as Marino and Merskin aptly observe, 'anthropomorphism is only anthropomorphism when a "human" trait does *not* actually exist in the other animal'.[69] The research in neuroscience, which we turn to now, provides a rich resource of data for the appreciation of the animal brain/mind, including dispositions for grief and spirituality.

[…] "I don't anthropomorphize," Frohoff told me. "I leave it to other people to do that. What I do is study gray whales using the same rigorous methodologies that have long been used to study the behaviors of other species and interspecies interaction. Those who would reject out of hand the idea that whales are intelligent enough to consciously interact with us haven't spent enough time around whales.'"

66   Safina 2013, 29.
67   de Waal 2019, 165.
68   e.g. Bradshaw and Sapolsky 2006.
69   Marino and Merskin 2019b, n.p.

From neurons to neighbours[70]

With greater insight into structural and functional characteristics of the brain came greater awareness of human–nonhuman animal brain/ mind comparability, which facilitates predictions of their subjective states. Expanded knowledge of the brain has also revealed the potency of affect and the implicit (outside reflective awareness) domain, and their primacy over explicit cognitive functions. Critically, both the experiences of grief and spirituality are supported by these implicit processes with cognitive elaboration playing a secondary role.

Primacy of ancient brain regions

In July 2012 a group of prominent neuroscientists, gathered in Cambridge, UK, at the Francis Crick Memorial Conference on Consciousness in Human and Nonhuman Animals, signed the Declaration on Consciousness,[71] which states that nonhuman animals are conscious beings. Specifically, the statement reads:

> The absence of a neocortex does not appear to preclude an organism from experiencing affective states. Convergent evidence indicates that non-human animals have the neuroanatomical, neurochemical, and neurophysiological substrates of conscious states along with the capacity to exhibit intentional behaviors. Consequently, the weight of evidence indicates that humans are not unique in possessing the neurological substrates that generate consciousness. Non-human animals, including all mammals and birds, and many other creatures, including octopuses, also possess these neurological substrates.

The Cambridge Declaration is a result of decades of more or less invasive research into the brains of various nonhuman animal species. Evidence from this research, contrary to traditional views, indicates the presence of phenomenal consciousness in subcortical regions of the mammalian

---

70 Cf. Shonkoff and Phillips 2000.
71 https://bit.ly/34Ea7BS

brain and corresponding areas of non-mammalian species. That is to say (using as an example the mammalian brain, which may be most familiar to the reader) that affective states are already experienced at the subcortical level without needing to be 'converted' into experience by higher and newer brain structures, namely the neocortex. Traditionally, as Panksepp, co-editor and co-signatory of the Cambridge Declaration, summarises, the neocortex was believed to be not only

> the seat of conscious thought, but also of emotional feelings [...] they [emotional feelings] were commonly deemed to be a form of thought, and affective and cognitive processes were envisioned to be completely interpenetrant in higher brain regions that generated certain higher cognitive processes such as frontal cortical regions.[72]

The obfuscation of thought and feeling and the belief that an emotional experience had to be processed through higher brain regions to acquire experiential relevance have had clear implications for nonhuman animals. Given that the neocortex can vary substantially among mammalian species,[73] many thinkers hesitated (and some still do) to attribute to other animals the capacity for subjective experiences, or to consider these experiences equivalent to those of humans. Just like ethologists, neuroscientists were wary of anthropomorphism, and this precluded discussions of animals' affective experiences in the field. By moving the loci of affect to subcortical regions, which are far more similar across species, the picture changes dramatically.

In the 1930s, Swiss physiologist and Nobel Prize-winner Walter Hess elicited rage in a cat by electrically stimulating her hypothalamus. Fearing that his work would be marginalised in the zeitgeist, Hess decided to label the responses as 'sham-rage', despite his conviction that those responses reflected true subjective experiences, as he admitted later in life.[74] Most other scientists followed suit, focusing on memory

---

72 Panksepp 2011, 3.
73 For a comparative picture, see Comparative mammalian brain collections at http://neurosciencelibrary.org.
74 Reported in Panksepp 2011.

and learning through observations of responses to stimuli, and studiously avoiding attributing feelings to their nonhuman experimental subjects. In this framework, emotions tend to be seen as behavioural and physiological responses that are not subjectively experienced – a kind of automatic, unfeeling instinct. Joseph LeDoux, for instance, who spent over thirty years studying the amygdala (an almond-shaped structure that is linked to fear responses, among other things) using fear conditioning[75] (which he prefers to call 'threat conditioning' as he does not believe animals experience fear[76]), reiterated again in 2012 that the question we should ask is 'to what extent are functions and circuits that are present in other mammals also present in humans' as opposed to 'anthropomorphically [...] asking whether human emotions/feelings have counterparts in other animals'.[77] It is worth noting here that the threat of what was considered as anthropomorphising (namely, attributing feelings to nonhuman animals) posed to these scientists as they were developing their careers (for example, the aforementioned warning in the journal *Science*) was in itself a form of fear conditioning.

Conditioning, however, can also be used with positive affects and effects. For instance, priming a place with a positive stimulus (for example, opiate administration) can lead to place preferences.[78] I return to place preferences/attachment in Chapter 4, where I propose that place attachment (or aversion) may play a role in the development of animals' (including humans') sensitivities for detection of and implicit communication with intangible agencies at the base of spiritual experiences.

While Hess himself did not pursue this line of research (the possibility of subjective experiences being generated in these ancient brain regions), others have. For instance, Jaak Panksepp, whose PhD

---

75   Classical or Pavlovian conditioning is a learning process in which a biologically significant stimulus (e.g. electric shock) is paired with a neutral stimulus (e.g. sound); if the pairing is repeated the previously neutral stimulus (sound) alone elicits a fear response.
76   The only feeling LeDoux believes nonhuman animals may experience is pain (reported in Panksepp 2010).
77   LeDoux 2012, 654.
78   Nelson and Panksepp 1998, 438; see also Prus, James and Rosecrans 2009.

dissertation focused on mapping the aggression system in rats for the first time, as self-reported in a *Brain Science Podcast* interview,[79] spent his working life researching these raw affects and became one of the most passionate advocates of the proposition that emotional feelings arise in those regions. These felt body states, as Damasio puts it, 'are also primordial and indispensable components of the self and constitute the very first and inchoate revelation to the mind that *its* organism is alive.'[80]

In essence, the weight of evidence suggests that self-relatedness (as *feeling* the self, not *thinking* it) cannot be considered as the output of higher-order tertiary processes, which, in mammals, are the domain of the neocortex. Instead, basic self-relatedness should be viewed as the input to the cognitive functions with their control and modulatory properties.[81] This cognitive modulation enables us, Han and Northoff explain, 'to distance ourselves from our own self' by taking an observational and analytical perspective instead of an experiential one,[82] but it does not per se generate experience.[83] Recent research points to the existence of a non-reflective, affective self-related processing, which underlies any kind of self-awareness, also referred to by some researchers as the 'core self'.[84]

In mammals, the core self is proposed as a coherent integrated system of self-representation with the epicentre in the subcortical–cortical midline systems.[85] It is postulated to have evolved from the 'proto self'[86] (a neural map of the body developed early in brain evolution, promoting coherence of the various functions) with the emergence of

---

79 Panksepp 2010.
80 Damasio 2010, 76.
81 Han and Northoff 2009, 204.
82 Han and Northoff 2009, 204.
83 'Whether the neocortex has *any* evolutionary-based affective functions, as opposed to learning-dependent development, is currently unresolved. It can surely engender a host of emotional thoughts and behaviours. Still, it seems, the epicentres for emotional affects remain subcortical, even though ancient cortical areas such as insula can generate various specific sensory affective feelings such as disgust and pain, but surely not without the participation of subcortical circuits' (Panksepp and Biven 2012, 451).
84 Panksepp 1998; Northoff et al. 2006; Northoff and Panksepp 2008.
85 Northoff and Panksepp 2008.
86 Damasio 1999.

primary-process emotional and motivational systems.[87] By combining rudimentary sensory impressions (sound, sight, etc.) and information from the body's interior (water levels, thermal properties, etc.) with emotional affects and action tendencies (autonomic responses, change in muscle tone, etc.), a sense of *mineness*, of ownership of the affective experience,[88] a basic self-relatedness of the organism, is determined in these ancient structures.[89] A water and energy imbalance, for instance, would result in the affective representation of thirst and hunger in the core self.[90] Subcortically generated self-relatedness appears to be characterised by immediacy, the here-and-now, dependent on the actual presence of interoceptive (emerging within an organism) or exteroceptive (external to an organism) stimuli, and the arousal of emotional systems and related responses.[91] Its connectivity to and interactivity with other parts of the brain determine its elaboration in cognitive and temporal terms. This enables self-representation without actual stimuli,[92] and the development of unique, idiographic selves based on the individual animal's experience in the early developmental context and all through life.

In this view of self-referential processing the neocortex is far from being just the icing on an already baked cake, as has been suggested by some critics.[93] Cortical regions, which grew not just on top of more ancient regions but also from and with them,[94] are deeply implicated through top-down[95] and bottom-up[96] modulation, contextualisation

---

87 Panksepp and Biven 2012, 394.
88 Panksepp and Biven 2012, 394, 400, 418, 419.
89 The most lucid, albeit ethically questionable, example of the subcortical system at work is provided by decorticated animals – animals who had their cortex surgically removed for research purposes. These animals appear to exhibit unimpaired emotional urges and behaviours, such as exploratory urges, playfulness, maternal care, lust, anger and fear (Panksepp and Biven 2012, 450). Human children born without the neocortex also retain consciousness (Merker 2007).
90 Panksepp and Biven 2012, 418.
91 Han and Northoff 2009, 207.
92 Han and Northoff 2009.
93 Feldman Barrett et al. 2007.
94 Damasio 2006 [1994], 128.
95 Cortical control of lower brain regions and physiology.
96 Feeding information from lower levels up.

and general enrichment of the subjective experience, not all of which is necessarily beneficial to the organism. As discussed in more detail in Chapter 2, the experiential matrix of an individual imprints memories and determines cognitive biases long before the brain is capable of episodic autobiographical memories and contextualisation of life events. This potentially leads to intense feelings and little understanding of their origins. The imprint of traumatic experiences can be particularly deep and long-lasting, leaving 'emotional systems sensitized or desensitized, with permanent, epigenetically induced high-stress reactivity and excessive primary-process negativistic feeling'.[97] An individual can quite literally get 'stuck' in a painful state of mind, exhaust all the available cognitive resources and still remain there. Complicated grief, mentioned earlier and discussed in more detail further on, is a case in point.

In his book *Mindsight*, Daniel Siegel, clinical professor of psychiatry at UCLA, offers several examples of patients with psychosomatic disorders caused by trauma or grief whom he has helped to heal. He stresses the importance of embracing an adverse feeling instead of fighting it, and appreciating the critical role the neural circuits that are currently sustaining the unfortunate mental state had in the history of brain evolution. Discussing a case of obsessive-compulsive disorder, Siegel reminds the reader, 'whether you're twelve or ninety-two, you're probably not going to win a battle against a brain circuit that's at least one hundred million years old. In an integrative approach, the winning strategy is respect and collaboration'.[98] His cognitive approach is complemented by body-focused techniques, including meditation, which, having originated in Asian traditions,[99] is increasingly being recognised as central to psycho-physical wellbeing and the restoration of such, and hence incorporated in healing practices worldwide.

Meditation has been shown to change brain function and structure with numerous beneficial outcomes for physical and mental health, including positive effects on stress responses, immune function, and

---

97 Panskepp and Biven 2012, 434.
98 Siegel 2011 [2010], 248.
99 See, for example, Samuel 2008.

even enhancement of natural reward processing in the brain, which may attenuate substance abuse.[100] By attending to the present moment, the narrative self with its cognitive elaborations gives way to the experiential self with the focus on moment-to-moment experience, which Farb and colleagues suggest operates via a distinct neural mode of self-reference, employing evolutionary older brain regions.[101] Silencing the cognitive self, the organism, as an intrinsically relational entity, enters into a dynamic albeit implicit communication with the world through interoceptive and exteroceptive cues. This may give rise to a felt sense of presence and produce a feeling of connection, of merging of the self with the nonself, opening a window on how spiritual experiences may arise in animals.

Psychologist Allan Schore, also from UCLA, explains that in the 1970s and 1980s science moved from observing (and attempting to correct) external maladaptive behaviour to focusing on internal cognitive processes.[102] Dominance of the cognitive paradigm, in human psychotherapy, resulted in the emergence of cognitive-behavioural therapy models, oriented towards changing an individual's conscious cognition; that is, their maladaptive thoughts and beliefs. The (re)integration of bodily based emotions into the equation following cumulative evidence of the inseparability of biological (body) and psychological (mind) processes is not only progressively informing the development of new therapeutic models for human conditions as well as preventive measures but, as Schore notes and as discussed above, also provides a solid foundation for the extension of such consideration to nonhuman animals and a more integrated view of their lives and being.

Ultimately, the animal brain is a highly interconnected, integrated system, while being at the same time very plastic, open to external influences and capable of internal reorganisation, for better or for worse. The implicit storeroom is continually updated throughout life, influencing feeling, thinking and decision-making, often creating values and attitudes without the individual's explicit consent.

---

100 e.g. Davidson et al. 2003; Davidson and McEwen 2012; Garland, Froelinger and Howard 2015.
101 Farb et al. 2007.
102 Schore 2012, 4.

Brain asymmetry

The shift in focus, in research and therapy, from cortical to subcortical regions mentioned above is accompanied also by a shift from the explicit left brain[103] and its language-based cognition to the implicit right brain and its dominance in emotional and corporeal processing.[104] A comprehensive presentation of the asymmetric brain, drawing from a wide range of research on the topic, with implications for our current Western way of life, is delivered in psychiatrist Iain McGilchrist's 2009 book, *The Master and His Emissary: the Divided Brain and the Making of the Western World.* The following paragraph summarises briefly the distinction between the two brain parts, based on the presentation in McGilchrist's book.

When we speak about brain lateralisation, we do need to keep in mind that both hemispheres are actively involved at any given time and there is probably nothing that is limited to one of them alone.[105] However, the hemispheres do appear to be specialised for attending to the world in different ways. The right hemisphere is implicit, global, holistic, processes novelty, and perceives things in their context.[106] The utilitarian left hemisphere, oriented towards the purpose of 'getting and feeding',[107] is explicit, reductionist, processes the familiar, and perceives things abstracted from the context. The left hemisphere subsequently reconstructs the broken-up parts into a different 'whole'. This new 'whole', a re-presentation, is a substantially simplified version of the wider, complex reality/ies of the world but it enables us a more or less successful 'manipulation' of the world and navigation in it. This manipulation, however, could not be achieved through the left hemisphere alone because the synthetisation of the 'two worlds' into a useable whole appears to be the sole capacity of the right hemisphere.[108]

---

103 The right/left hemisphere describes the cerebral cortical areas, while the right/left brain includes cortical and subcortical structures.
104 Schore 2005, 205; for a review see Gainotti 2012.
105 McGilchrist 2009, 10–11.
106 For a more comprehensive presentation see McGilchrist 2009, especially Chapters 1 and 2.
107 McGilchrist 2009, 27.
108 McGilchrist 2009, 176.

At the same time, due to the exploratory, global nature of the right hemisphere and the narrow focus of the left hemisphere, the left hemisphere's attendance to the world is biased from the beginning, in that the left hemisphere is dependent on and limited by the attentional context the right hemisphere had prioritised.[109] In sum, the right hemisphere needs the left, but the left hemisphere is dependent on the right; the latter process is literally vital, the previous is not.[110]

Functional brain lateralisation is a cross-species phenomenon, present also in invertebrate animals, such as, for example, spiders and ants.[111] Without denying some species-specific peculiarities,[112] like in humans, in other animals the left brain also tends to be specialised for operations of getting and feeding, while the right brain processes global information and facilitates socio-emotional interactions, including attachment processes. Chicks, for example, use the right eye (and by implication the contralateral – left – hemisphere) to distinguish between seeds and background during feeding, while keeping the left eye (right hemisphere) engaged in monitoring the rest of the environment. When chicks use their left eye (right hemisphere) to peck between grains and pebbles, they peck randomly.[113] For attachment ('imprinting') memories on the other hand, it is the right brain that is

---

109 McGilchrist 2009, 44.
110 McGilchrist 2009, 219, 227.
111 Heuts and Brunt 2005, 601–2.
112 Vallortigara and Rogers (2005) report, for example, that the European starling appears to perform food-discrimination tasks better using the left eye (right hemisphere), but also point out that such species variation does not necessarily testify to differences in laterality; rather it could be a reflection of a different behavioural strategy, e.g. relying more on spatial than non-spatial clues during foraging (p. 577).
113 Reported in Vallortigara and Rogers 2005; see also Buschmann, Manns and Güntürkün 2006. This left-hemisphere capacity appears to develop in the last stages of the embryo development, just before hatching, when in natural conditions the embryo of most avian species would turn around in the egg, exposing the right eye (left hemisphere) to the light entering the eggshell. The impairment of this type of lateralisation, e.g. by incubating the eggs in the dark, also affects the social hierarchy among young chicks: normally lateralised chicks establish a more solid hierarchy compared to impaired (hatched in the dark) chicks.

critical[114] as it is for their ability to discriminate between strangers and companions.[115] Similarly, monkeys[116] and sheep[117] have been shown to use the right hemisphere in the recognition of individual faces, and, in the case of monkeys, also of facial expressions. Whale calves – as well as bottlenose dolphin calves – display a clear left body side / right hemisphere bias in relation to their mothers, and later in relation to other conspecifics, by choosing to keep the mother in their left visual hemifield (one half of the visual field)[118] whereas ants have been observed preferentially lying down with their left body side exposed to a nest-mate.[119]

Such right-hemisphere bias finds curious manifestations in human art. A review of a body of sculptures and paintings depicting mothers holding their infants and dating as far back as 2000BCE unveils a clear left-side cradling bias – the infant's head lying on the left side of the mother's body, thus leaving the infant's left eye, as well as left ear, more exposed to the mother.[120] Real-life observations of humans also reveal such left-side cradling bias as do observations of chimpanzees and gorillas.[121]

---

114 Bolhios and Honey 1998; Moorman and Nicol 2015. While these memories are stored in both the left and the right brain, the left brain stores a fixed representation of the imprinted stimulus, while the right brain is more flexible and holistic, capable of detecting small changes and updating the memories with new information.

115 Vallortigara and Andrew 1994.

116 Hamilton and Vermeire 1988.

117 Peirce, Leigh and Kendrick 2000.

118 Karenina et al. 2010.

119 Heuts and Brunt 2005, 602.

120 Grüsser1983; Grüsser, Selke and Zynda 1988.

121 e.g. Harris et al. 2001; Alzahrani 2012 for humans; Manning, Heaton and Chamberlain 1994 for chimpanzees and gorillas. As Lori Marino (email, September 2019) notes, there could be other factors influencing this left-side cradling, such as the left-side position of the heart and babies being soothed by heartbeat, and with most humans being right-handed they may prefer to use the right hand for more delicate manipulations, including stroking the baby.

Facilitated by the right brain, emotional communication[122] and learning, along with implicit affective memories,[123] play a critical role in attachment formation (and loss as a result of such attachment) as well as in spiritual relating to other agencies. As an intrinsically relational entity, from conception onward, the organism is in continual communication with the environment. This communication shapes the organism's internal fabric. Relationality begins in the womb (and the egg), which is not a vacuum but a dynamic environment with sensory input reaching the foetus and the foetus responding to it. The study of chicks, for instance, suggests that they are able to communicate with chicks in other eggs and synchronise hatching,[124] while for human babies it has been shown that the sounds the foetus is exposed to influence the baby's (post-natal) vowel perception.[125] The post-natal period continues to be critical for many animal infants of both precocial and altricial species.[126] Apart from physical security and food provision, the primary caregiver (usually the mother but not always) also acts as an external regulator of the infant's developing capacity for self-regulation. This is to say that the infant's maturing self-regulatory system grows and develops in interaction with the caregiver. The quality of such interaction, this right-brain-to-right-brain communication, as Schore[127] terms it, will determine the efficacy of self-regulation with lifelong implications, including response to grief. Through these interactions internal patterns develop – I elaborate on this in Chapter 2 – entirely outside conscious awareness, on an implicit experiential level. As such, they may also remain inaccessible to the

---

122 This includes verbal emotional communication: while the left hemisphere is dominant in processing the semantic content (the 'what') of the emotional message, prosody (the 'how') falls into the domain of the right hemisphere (e.g. Vingerhoets, Berckmoes and Stroobant 2003).
123 e.g. Schore 2011; Mancia 2006; Gainotti 2012.
124 Broom and Fraser 2007.
125 Moon, Lagercrantz and Kuhl 2013.
126 Precocial animals are relatively independent soon after birth, able to walk, feed themselves, flee, e.g. deer and ducks; altricial animals are relatively helpless and immobile at birth and completely dependent on adults for food, e.g. humans and dogs.
127 Schore 2011.

reflective self, yet alive and well, for better or for worse, on a purely experiential level.

The vulnerability of the BrainMind[128] and the limited power animals sometimes have to protect ourselves from being indelibly marked by adverse influences can be frightening. Nevertheless, not everything that touches us is pathogenic. The propensity for relationality and the intrinsic sensitivity to intra- and inter-organismic exchanges also enables positive affect: the pain of grief is a consequence of the delight of love. The implicit mechanisms and memories that can lock us into a place of agony and feelings of no-escape parallel those that open us up to the rest of being, enabling, among other things, the 'dance with animacy', which I will put at the core of spiritual experience, as detailed in Chapter 4.

As a slight diversion from functional asymmetry, it is relevant to briefly explore another asymmetry, which works to the advantage of the left hemisphere and enables its point of view to become dominant[129] despite the right hemisphere's overall primacy, namely what McGilchrist calls the 'asymmetry of means'.[130] The left hemisphere's nature of re-presenting reality in an ultimately reductionist way, its capacity to build abstract (and abstracted) categorisation systems, its favouring of analytic, sequential processing and the construction of systematic thought based on apparent certainties, combined with its control of syntax and lexicon, can result in a substantial shift of balance. McGilchrist argues, and I agree, that to a great extent our current mechanistically oriented world, including our insensitivity to other animals, is a reflection of this shift. The tacit right hemisphere with its parallel processing of information has neither the linearity needed to construct an argument that could persuade nor a voice to express it, while the left hemisphere has both.[131]

---

128 Term used to indicate brain–mind connectivity and its monistic nature in contrast to the dualistic view of brain–mind separation.
129 At least in some human societies with obvious impact on the rest of the world.
130 McGilchrist 2009, 228–9.
131 Though lacking explicit verbal properties, the implicit domain is hardly mute. Intersubjective nonverbal implicit communication, discussed above, plays a crucial role in our early formative period and accompanies us throughout life.

When it comes to nonhuman animals, a substantial amount of prejudice against them, including humans questioning other animals' capacities for thinking, results from their supposed lack of a verbal language. However, thought does not appear to be dependent on language, although given that meta-thinking – the conscious introspection of our thought processes – does employ language, we might be misled into believing that language is somehow necessary for thought. Most of our thinking and sense-making of the world, category and concept formation (which I discuss in more detail shortly), evaluation of evidence, and problem solving is achieved without language and mostly without being consciously aware of it.[132] Aphasic patients, who have lost the ability of speech, might provide immediate evidence of non-linguistic thinking. J. Lordat, a professor of physiology, noted following recovery from his aphasic episode: 'when I wanted to speak I could not find the expressions that I needed ... *the thought was all ready*, but the sounds that had to express it as intermediary were no longer at my disposition.'[133] We might be able to recall times when during a speech, or even during writing, we ourselves came to a point where a word was nowhere to be found. The concept, the thought, was there, but in that instant the word, the verbal representation of the

---

Furthermore, while speech arises in the left hemisphere in humans, as do instrumental vocalisations in other species (Corballis 2002, reported in McGilchrist 2009, 27), the right hemisphere appears to have a much larger vocabulary than traditionally thought (Querné, Eustache and Faure 2000). The right hemisphere's vocabulary, however, consists of long, unusual words, mirroring its nature of non-focal information processing and a broad semantic-associative field, as opposed to the left hemisphere's predilection for close lexical semantic connections with the tendency of suppressing associations that are currently not relevant (McGilchrist 2009, 41, 51). Nevertheless, in the case of left hemispherectomy (surgical removal of the left hemisphere), the right hemisphere may be able to substantially compensate for the missing left (e.g. Vanlancker-Sidtis 2004, and Smith-Conway et al. 2012). Besides, comprehension of the meaning of sentences and phrases in context, the tone, emotional power, metaphor, irony and similar are subserved by the right hemisphere (McGilchrist 2009, 99); see also footnote 122.

132 McGilchrist 2009, 107; Snyder, Bossomaier and Mitchell 2004.
133 Cited in McGilchrist 2009, 109.

thought, was not. As a consequence of his experience, Lordat came to the conclusion that the theory which posits verbal signs are necessary, even indispensable for thinking, could not possibly hold.

With human verbal language held in such high regard and with nonhuman animals' inability to articulate their feelings and thoughts in such a language, it may be tempting for some humans to assume that there are no feelings and thoughts in other animals, or that their feelings and thoughts are somehow of a lesser intensity and value. However, it is becoming increasingly evident that not only are nonhuman animals capable of intentional and instrumental communication, but their vocabulary can be impressively broad. For instance, prairie dogs, extensively studied by Con Slobodchikoff from Northern Arizona University, have a sophisticated language, which includes, among other things, social chatter and vocalisations for distinguishing triangular from circular shapes.[134] Hens have specific vocalisations to distinguish aerial predators from threats on the ground.[135] Dolphins call one another by name.[136] Cowbirds produce structurally different songs based on different developmental and learning contexts.[137] Nightingales have elaborated songs in which each note is sung in a one-tenth of a second, making it difficult for humans to appreciate the nuances unless the song is recorded and slowed down for replay.[138] What to us might sound like a chirp could in fact be conveying a full balladic stanza to the informed listener; that is, another nightingale. These and other known or not-yet-discovered communication skills leave little doubt that the nonhuman animal world is also replete with vibrant information exchange – a topic that has more recently been taken up by Dutch philosopher Eva Meijer.[139]

Language and communication aside, just like in the case of emotions, thought generation and processing mechanisms in nonhuman animals parallel our own. In a 2008 essay, neuroscientist

---

134 Garber 2013, n.p.; see also Slobodchikoff 2012.
135 Reported in Potts 2012, 45.
136 King and Janik 2013.
137 West, King and White 2003.
138 Reported in Balcombe 2010, 21.
139 e.g. Meijer 2019 [2016].

Giorgio Vallortigara and colleagues dispute the idea that nonhuman animals' cognition resembles that of human autistic savants, proposed by Temple Grandin, herself an autistic savant and advocate of the oxymoronic 'humane slaughter' of nonhuman animals.[140] This idea has also been supported by some neuroscientists who are not experts in animal cognition, as the authors of the above essay explain. Autistic people have direct access to raw sensory information, hence paying attention to details instead of the whole picture.[141] A 'normal' animal (human inclusive) brain, on the other hand, tends to process sensory data through acquired concepts, whereby the term 'concept' describes groupings of attributes, sensory details, characteristic of familiar 'objects', while 'object' is intended in the broadest sense and it includes not only tables and chairs but also complex interactions and events.[142] Following the formation of a concept, the details constituting the concept elude conscious awareness in lieu of the bigger picture. While this process is necessarily reductionist in nature, it is very useful for animals; without the capacity for such concept formation and categorisation of phenomena and situations, we would have to start anew with every detail that hits our senses. Animals would be overwhelmed by details and the efficiency of navigation and decision-making in the socio-natural environment would be significantly compromised. Concept formation appears to be a specialisation of the left hemisphere, while the gestalt-oriented right hemisphere continues to search for novelty and to challenge the established templates. 'Overall, and very generally,' as Vallortigara and colleagues put it, referring to both human and nonhuman animals,

> the left hemisphere sets up rules based on experience, and the right hemisphere avoids rules in order to detect details and unique features that allow it to decide what is familiar and what is novel.[143]

---

140 Grandin 2005.
141 An autistic boy, for instance, learnt the concept of giraffe through the coat pattern not the overall animal, later identifying a leopard as a giraffe (reported in Vallortigara et al. 2008, 0210).
142 Snyder, Bossomaier and Mitchell 2004, 34.
143 Vallortigara et al. 2008, 0213.

For the overall discussion in this book, an insight into concept formation is relevant for two principal reasons. First, the highly automatic conceptualisation and categorisation system is vulnerable to the generation of illusion and prejudice. The function of this system is to offer the best possible hypothesis based on experience; this makes us see what we expect and know, not necessarily what we may actually be experiencing, and we tend to fit stimuli into known concepts.[144] In relation to nonhuman animals, most of us in Western society were born into a familial environment in which other animals' inferiority and humans' 'rights' to use them were taken for granted. Our mental templates developed accordingly. We were never given a choice, and neither were our parents, or our teachers. These templates may be resistant to change; there is safety in familiarity and novelty is unpredictable, which aids understanding people's tendency to engage in system justification, mentioned earlier. Nevertheless, once we recognise that there was little volition and conscious thought involved in the development of our views of other animals (and the world more generally), we also recognise the opportunity for re-examination of our attitudes and the formation of a more considerate and deliberate choice (regardless of what this turns out to be).

'The nature of attention one brings to bear on anything alters what one finds,' writes McGilchrist. 'To attempt to detach oneself entirely is just to bring a special kind of attention to bear which will have important consequences for what we find.'[145] Science and its presumed objectivity as well as philosophy and its presumed rationality are equally value-laden forms of attentiveness, other ways of looking at things, chosen premises upon which neutral cognitive functions build a truth, a truth that, unlike the cognitive functions, is in itself never neutral, always compromised.[146] Fact and fetish (belief) work hand in hand, and a substantial part of scientific ('objective') investigations into nonhuman animals' existence in the past hundred years has consisted of what French philosopher Bruno Latour (not referring to nonhuman animals) would call isolated scientific images. When extracted from the

---

144 Snyder, Bossomaier and Mitchell 2004; Vallortigara et al. 2008.
145 McGilchrist 2009, 29.
146 McGilchrist 2009, 28.

chain of images, such an image 'has no truth value, although it might trigger [...] a sort of shadow referent, which is taken, by a sort of optical illusion, to be the model of the copy'.[147] The second reason concept formation is of interest here is its role in the experience of awe, and spirituality more broadly, as we will see in Chapter 4. Awe has been proposed to emerge at the encounter of a phenomenon that does not fit into an established mental template and that the brain thus cannot automatically assimilate to achieve cognitive closure.[148] Since other animals also process information using mental schemas, it is likely that they too encounter phenomena that the brain cannot automatically assign to a known category, hence generating a sense of awe. The existence of individual differences in the need for such closure[149] within the human population is also a factor worthy of consideration, particularly since the cultural context may play a role in the shaping of uncertainty tolerance or intolerance.

This introduces the interesting question of how tolerant (or not) nonhuman animals may be in relation to uncertainty, and how open (or not) they may be to the relational potentials of these encounters and as a consequence to spiritual experiences. I will look for answers principally in the following three areas: a) organismic capacities for communication on an implicit, non-reflective, experiential level, as outlined above; b) experiential focus ('staying in the present moment') versus narrative focus, and how prioritising one or the other may impact upon such communication; and c) the role different self-construals (independent, interdependent/ecocentric) may play in facilitating or inhibiting these relations.

While I do not delve into Western philosophy in any notable detail, it is necessary to acknowledge the mental effort, particularly over the past century, within this philosophical tradition to bridge the mind–body gap from the past. Parallel to the growing interest in and understanding of embodiment in other disciplines, philosophy has also been moving away from dualism towards an intersubjectivity grounded in relationality, both from a developmental perspective and

---

147 Latour 2010, 114.
148 Valdesolo and Graham 2014.
149 Webster and Kruglansky 1994.

post-developmental enactment. Maurice Merleau-Ponty,[150] in particular, has influenced thinkers with his emphasis on the primacy of embodiment along with the recognition of the organism's capacity for 'a communication with the world more ancient than thought',[151] and his criticism of philosophical streams, both idealist and realist, that fail to take into account the embodied nature of perception: 'man is in the world, and only in the world he knows himself'.[152] Building on the work of philosophers Merleau-Ponty, Edmund Husserl, Martin Buber and others, Nick Crossley,[153] for instance, discusses the concepts of radical intersubjectivity and pre-reflective consciousness, which (even though he remains unnecessarily anthropocentric) resonate with notions of the experiential self and cross-species relationality and relationability addressed in this book. An important observation Crossley makes is that 'human subjectivity is not, in essence, a private "inner world"' separate from the outside world, instead it consists of 'sensuous, embodied beings and [...] it is therefore public and intersubjective.'[154] The same may be said to apply to other animals. Could their subjectivity be more public than we have been willing to recognise?

Integration

The above brief overview provides an insight into the implicit forces at work in the shaping of animals' internal landscapes. From neurons to neighbours, we become who we hang out with, and often, such as in the early formative period, we have little choice over it. However, the experiential matrix of an individual continues to frame and reframe feelings and thoughts, which guide animals through 'the splendour and travail' of life.[155] Cognitive appraisal and interpretative outcomes of phenomena and events, which, just like humans, other animals also engage in, may differ substantially among animal species and are perhaps more difficult to assess in nonhuman animals. The structural and

---

150 Merleau-Ponty 1963 [1942]; Merleau-Ponty 1962 [1945].
151 Merleau-Ponty 1962 [1945], 296.
152 Merleau-Ponty 1962 [1945], xii.
153 Crossley 1996.
154 Crossley 1996, 24.
155 Beston 1928, 25.

functional properties underlying animals' experiential consciousness, including those generating grief and spiritual experiences, however, share commonalities that enable discussion of subjectivity with high probability predictions.

Attachment theory and grief

Overview

To be able to appreciate the affective impact of the loss of a proximal subject, we need to first understand principles of intersubjective attachment. In the 1950s when John Bowlby (introduced at the beginning of this chapter) began work in the area of infant–caregiver separation, he had, by his own admission, 'no conception of what [he] was undertaking'; the subject appeared to him 'a limited one, namely, to discuss the theoretical implication of some observations of how young children respond to temporary loss of mother'.[156] The negative impact of prolonged separation (such as in the case of hospitalisation of children) was clear, the hows and whys less so. It was only upon familiarisation with the ethological work of Konrad Lorenz and Niko Tinbergen, particularly the phenomenon of imprinting in precocial avian species, that Bowlby was able to envision the possibility that mammalian (including human) attachments may work in a similar fashion – a view contrary to the then contemporary theories of human attachment. By converging fields pioneered by his two most significant intellectual influences, Darwin and Freud, Bowlby, in the words of his colleague Mary Ainsworth, in essence attempted to 'update psychoanalytic theory in the light of recent advances in biology'.[157]

In Chapter 2 we will examine more closely the context in which Bowlby's theory developed, and how field evidence did not support existing theoretical perspectives on infant attachment centred around food acquisition. Specifically, socially deprived children (for example, those in orphanages) were not thriving emotionally and cognitively

---

156 Bowlby 1982 [1969], xxvii.
157 Cited in Schore 1982, xi.

despite high standards of physical care and nutrition.[158] Harry Harlow's invasive experiments on rhesus monkeys provided further evidence of the negative impacts of early social deprivation and its cross-generational transfer. All this indicated that attachment relations and social bonds play a far more critical – vital – role than had been imagined. Research over the following decades has confirmed this supposition, demonstrating the crucial role of the primary caregiver for the infant's psychobiological regulation: the primary caregiver effectively serves as a regulator to the infant's developing self-regulatory system that will accompany the animal throughout life.

This early period is critical also for the development of the infant's sense of self. '[W]e first know ourselves as reflected in the other,'[159] and we begin acquiring this 'knowledge' before we are consciously (reflectively) aware of ourselves. The quality of the interaction in the infant–caregiver dyad influences attachment styles and internal working models (mental representations of the self and the world) that inform animals' future relations and can also impact upon our grief responses. The notion of secure and insecure attachment styles, discussed in the next chapter, was Mary Ainsworth's contribution to attachment theory. Ainsworth worked closely with Bowlby, providing empirical support for his theoretical frameworks.

Another important working relationship was forged in the 1960s in the attachment/loss arena, and that was the relationship between Bowlby and Colin Murray Parkes. Parkes, who continued his career in grief studies, had joined Bowlby's research unit in 1962, intrigued by Bowlby's work on infants' grief as a possible source of insight into the grief of adults. That was also when studies in adult attachment began, with Bowlby and Parkes focusing on adult bereavement,[160] and Robert Weiss examining marital separation.[161] These were followed by Hazan and Shaver, who concentrated on adult romantic relationships, and others. Elisabeth Kübler-Ross's influential book *On Death and Dying* was also influenced by Bowlby and his team's work, as was Cicely

---

158 Spitz 1945.
159 Siegel 2011 [2010], 62.
160 Bowlby and Parkes 1970; Parkes 1972.
161 Weiss 1973; 1977.

Saunders, the founder of the modern hospice movement, and the programs she and Parkes developed for the emotional care of dying and bereaved people.[162]

Of particular interest for the discussion of animal grief is Parkes's observation that one of the problems of discussing and analysing grief, which still exists today, is the lack of an accepted definition.[163] Most people define grief as the reaction to bereavement in a broad sense, and the 'measurements' include variables such as anger, self-reproach, depression, threats to our security, major changes in life, shame and guilt for our complicity or neglect that led to the death, etc. Parkes believes that while these potential realities (to which we should add the fear of mortality by creatures who believe they are mortal) might complicate grief and cause lasting consequences, they are not part of grief itself and might occur following a number of other stressful events and circumstances in life.

Definitions

In agreement with Parkes, *grief* in this discussion will be defined as: 'the experience of a loss and a reaction of intense pining and yearning for the object lost (separation anxiety). Without these a person cannot really be said to be grieving.'[164]

Along the lines of the definitions Mark Scott adopts in his essay 'Journeys in Grief: Theorizing Mourning Rituals', *bereavement* will denote the state of loss,[165] inclusive of side effects on an individual's life (for example, the need for internal and external reorganisation, etc.)

---

162 Reported in Bretherton 1992.
163 Parkes 2009 [2006], 29–30.
164 Parkes 2009 [2006], 30.
165 Scott 2009, citing Hockey, Katz and Small 2001, 4–5; and Corr, Nabe and Corr 2006, 205: 'The term *bereavement* refers to the state of being bereaved or deprived of something. In other words, bereavement identifies the objective situation of individuals who have experienced a loss of some person or thing that they valued. Three elements are essential in all bereavement: (1) a relationship or attachment with some person or thing that is valued; (2) the loss – ending, termination, separation – of that relationship; and (3) an individual who is deprived of the valued person or thing by the loss.'

after the loss of the attachment figure, which trigger, or at least call for, coping mechanisms. On the other hand, *mourning* will describe an outward expression of loss, the process of coping/dealing with it and the manner of incorporation of such processes into reality by individuals and/or societies.[166]

Current context

Much research has been conducted in human grief alone over the past few decades, yet many unanswered questions still remain, particularly in relation to prolonged grief, which some researchers advocated unsuccessfully be included as a separate diagnostic category of psychological disorders in the 2013 revision of the *Diagnostic and Statistical Manual of Mental Disorders*. Prolonged/complicated grief can affect human and nonhuman animals alike and will be discussed in more detail later on.

Nonhuman animal grief is a more recent area of investigation but interest in the topic is rapidly growing. This is reflected, for instance, in two recently published books, David Alderton's 2011 *Animal Grief: How Animals Mourn*, a concise, coffee-table-style overview of the knowns, unknowns and potentials of this field of inquiry, and Barbara J. King's more comprehensive 2013 volume, *How Animals Grieve*, a book rich with examples of grief expression across many species and situations, which are often accompanied by useful commentary, but with some definitional inconsistency. The experience of grief and the expression of grief are often used interchangeably,[167] and while a cross-species comparability of grief as an experienced emotion appears to be implied in the book, it is not explored in its own right but tends to be coalesced with behaviour and to some extent even with

---

166 Scott 2009, 81.
167 For instance: apart from showing that nonhuman animals can grieve, King also wishes to 'honor human uniqueness. Anthropologists have described many ways in which our species is unique in how we grieve,' and then she continues: 'Goat grief, then is not chicken grief. And chicken grief is not chimpanzee grief or elephant or human grief' (p. 7), and 'Why shouldn't our [human] grief be different? Evolutionary theory predicts species-specific behaviors in each animal' (p. 147).

cognitive capacities humans attribute to various animal species.[168] Even though King admits that feeling can exist without expression,[169] and emphasises that grief ensues from feeling rather than thinking,[170] the reader may nevertheless be left wondering whether other animals are indeed capable of experiencing human-comparable grief – a doubt that may be further fuelled by the presence of unexplained assumptions, such as that humans 'feel differently from other creatures [...] Why shouldn't our grief be different?',[171] and claims, such as: 'Do animals dwell on their sadness, closing their eyes at night aware that the blanket of grief will still be there at dawn? Probably not.'[172] This assertion may be valid but its function in this context is unclear and it has potential to exacerbate the human–nonhuman animal divide. The feeling is there now and it hurts. Do animals know it is going to go away? Probably not (and it may not go).

The growing interest in grief and death-related questions is also reflected in testimonies by veterinary practitioners whose help is being sought for grieving companion animals – one such case is discussed further on in this book – as well as in the emergence of the new field of evolutionary thanatology, launched in September 2018 through the publication of a themed issue of *Philosophical Transactions of the Royal Society*. While grief, as announced in the introduction[173] to the issue, will be one of the principal concerns of the field, at this stage the focus

---

168 On several occasions King brings up the question of mental capacities and speculates about various species' capacities for self-awareness and even traumatic memories (e.g. speaking of goats, p. 5: 'they [probably] don't recall past events or experience traumatic memories to the degree that elephants do. Their self-awareness is not as developed', and p. 147: a goat is 'less able to reflect upon her own life'). These speculations seem unnecessary; nonetheless, they carry implicit messages that may affect the way the reader views cross-species comparability of grief as an emotion; further, on p. 159 King writes that 100,000 years ago 'our ancestors had the cognitive and emotional resources to feel grief' without explaining what the minimum requirements for such capacities may be.

169 King 2013, 159.

170 King 2013, 14.

171 King 2013, 147.

172 King 2013, 148.

173 Anderson, Biro and Pettitt 2018.

appears to be of a behavioural-cognitive nature, specifically the collection of data concerning responses to death in nonhuman animals and the theorisation of possible understandings of death outside the human species. The authors in this collection continue to be wary of attributing human-comparable emotions (including grief) to nonhuman animals, but the initiative deserves applause for its willingness to take nonhuman animal death and the potential for grief seriously and for encouraging other scientists to come on board and become more attentive to death-related occurrences in other animals.[174]

The present volume is offered as a step towards a more comprehensive appreciation of the building blocks of animal (human inclusive) subjectivity in relation to loss, and to some extent spirituality. It could be viewed as complementary to existent works on the topic and those still in progress. Contrary to most other views, explicit or implied, the premise of this book is that grief – the internal reaction to loss – and its intensity may differ on an individual level (based on the individual's psychological disposition, attachment styles, circumstances, the nature of the relationship with the lost subject, etc.) not on a species level.[175] The principal gaps in the consideration of grief addressed here are: (i) the apparent limited appreciation of the potency of psychobiological phenomena underpinning separation anxiety and loss, and (ii) the lack of integration of human cross-cultural external manifestations of grief, including the absence of such manifestations, which calls for cautiousness in dismissing grief in nonhuman animals if explicit behavioural indicators are not present.

In Chapter 3, 'Cross-cultural grief matters' (pun intended), I examine selected human cultures whose death and mourning rituals depart significantly from those of the WEIRD, an acronym for Western, Educated, Industrialised, Rich and Democratic.[176] From stoic societies where outer expression of grief is not condoned, to those in which such expression is expected; from societies that bury their dead to those that do not; from societies that allow little or no interaction with the corpse to those that encourage physical proximity, offer food to the corpse, or

---

174 e.g. Watson and Matsuzawa 2018.
175 At least when it comes to mammals and birds.
176 Henrich, Heine and Norenzayan 2010.

even store the corpse in barrels used for rice wine; the diversity of grief expression and repression within the human species itself is a further signal that behaviour may be a poor indicator of individuals' internal states. It also suggests that, as in the case of expressing/repressing physical pain, other animals may have evolved species-specific and possibly also community-specific (while always allowing for variance on an individual level) ways of behaving around corpses that we may not identify as a manifestation of grief but the animal may be grieving nonetheless. The human cross-cultural context also offers the opportunity to tackle the question of 'understanding' death, which often arises in discussions of nonhuman animal grief. Considerations of agency detection and its vital role in human and nonhuman animals' lives suggest that animals are equipped with various perceptive tools that help distinguish between organisms that are dead and those that are alive.[177]

Spirituality and animals

The topic of spirituality could probably not have been further removed from my circle of concern when I began research in grief. This, however, changed following my four months living as a mother surrogate to a lamb who had come to us at one week of age after his biological mother's tragic death at a university farm. Wanting to offer him a normative upbringing (as much as the circumstances allowed it) I spent most of the time in the paddock with him and our two other rescued adult male sheep. My presence functioned as a secure base as predicated by attachment theory, while the other two sheep were effectively and affectively teaching him how to be a sheep. In the openness of the paddock and the adjacent forest area, the world became alive in a way that I had not experienced before. The sheep and other nonhuman residents were perfectly attuned to the environment. Every sound, every sight, every scent bore a meaning; the world was filled with agency speaking to them, and they, in their turn, were responding to it. Their moment-to-moment alertness (a stick could be a snake, a

---

177 e.g. Barrett and Behne 2005; Guthrie 1980.

fallen branch a predator roaming through the bushes) was expected and therefore not surprising, nor was surprising the rather automatic appraisal that followed disturbances – the evaluation of the nature and danger levels of single disturbances in accordance with internal categorisation systems discussed above. Could other animals, I pondered, encounter phenomena that resist categorisation, a condition that has been proposed to give rise to the feeling of awe in humans, potentially leading to either dread or bliss?

The wind, for instance, could obstruct such categorisation, possibly causing the feeling of dread. In the wind sensory cues become confusing, alertness may turn into vigilance, vigilance into anxiety, the sympathetic nervous system ('fight or flight response') may become aroused as agencies are felt to be all around but it is harder to detect and evaluate singular ones. On the other end of the spectrum, when nonhuman animals are resting, quiet and seemingly content, absorbing the warmth of the sun or sheltering in the shade of a tree on a hot day, immersed in the space and their own bodies, fully aware of themselves at the fundamental, experiential level, could they not be having what we may call a spiritual experience?

Nonhuman animal religiosity has received some scholarly attention over the past few years.[178] My view resonates with Donovan O. Schaefer's conceptualisations of religion as well as with new animism.[179] However, religion is a complex phenomenon consisting of implicit and explicit, interpretative aspects, whereas my focus is the intimate, implicit communication with that which is or becomes animated – a communication facilitated by the non-reflective, experiential consciousness, which relies on more ancient and implicit organismic properties and organismic relational potentialities that humans share with other animals.

In this context I consider the possibility of place attachment as the developmental origins of spiritual relating. Research in place attachment has focused more on social aspects of place: place as a container for social interactions,[180] rather than the direct relationship

---

178 e.g. Harrod 2011, 2014, 2016; Schaefer 2012, 2015.
179 Harvey 2005.
180 Lewicka 2011.

person–place, which will be of principal concern in discussing spirituality. We live and breathe spaces, we are attracted to some but not others, some may promote wellbeing, others not. From the womb (or egg) on, animals communicate, on an implicit level, with intangible agencies that constitute a space. Attachment figures (parents, partners, etc.) could themselves be viewed as a place with the multimodal support they offer, thus blurring the lines between subject and place bidirectionally. We will explore these issues by including in the discussion works that have examined this more intimate aspect of person–place relations, and ecopsychology. They provide useful insight into the propensity for and healing properties of connectivity with the rest of life (ecocentric self) as opposed to detachment (egocentric self), and into the devastating effects of environmental destruction on humans but particularly on other animals, bringing about 'solastalgia'[181] or profound *grief* for a place that used to be called home.

* * *

From subject to place, from human to other animals, where there is loss there is love, and (often) vice versa. The following pages will explore in more detail various aspects of these dynamic relations that mark the lives of humans and other animal nations. Its analysis will remain within the confines of conventional social and natural sciences. As much as it would be interesting to add a less 'local' perspective to the exploration of the vibrancy of existence and its intrinsic relationality, I have neither the space nor the expertise to do so. Nevertheless, it is worth keeping in mind that a biosystem, embedded in life and the product of evolutionary and developmental processes, may possess knowledge that may be inaccessible to science, as Josephson and Pallikari-Viras point out in their discussion of direct action at a distance (for example, telepathy);[182] after all, '[w]e are all connected ... made of star stuff'.[183]

---

181 Albrecht 2005.
182 Josephson and Pallikari-Viras 1991.
183 This is a short extract from a music video created by John D. Boswell, featuring scientists Carl Sagan, Richard Feynman, Neil deGrasse Tyson and Bill Nye. Further information and the video itself are available at: https://bit.ly/30JaLgs

2
Intersubjective attachment and loss

Grief is the price we pay for love.
Without attachment there would be no sense
of loss.
— Author unknown[1]

Bond brokering

We were fortunate, my partner and I, as we were about to waste a substantial sum on a 'shoebox' home in the inner city where we were renting at the time, to have been blessed with a moment of sanity, as we now see it, and looked outside the *box* our contemporary cultural niche was imposing upon us. Instead, we purchased a larger house with a garden in a small town in the mountains, at a manageable distance from the city. Three years later we moved to the edge of that same town onto a larger block of land and have lived here with other animal residents ever since. Over the years we have had the privilege to observe, and at times partake in, social interactions of various native and non-native species, including infant rearing. The less appealing but equally informative aspect of this place was its proximity to a small-scale facility that bred

---

1   Cited in Mallon 2008, 4.

sheep for their flesh. I recorded the following note during the first few months of moving here:

Today the two ducklings took their first flight, or at least the first flight that we, the human residents, have observed. It was a very short flight, about one metre long. For weeks now, we have been deeply preoccupied with the Australian wood duck culture. There are several wood ducks living on the property, including the family of four. The parents have over the weeks gained our deepest respect and admiration for the devotion they have exhibited for their new role as carers and educators of the young. The caregivers appear to be extremely protective of the young: they do not allow other ducks and other animals, including humans, dogs, sheep and rabbits, to come too close to the family nucleus. As soon as a foreign animal, that is, an animal who is not part of the family, reaches the proximity range of about two metres, the female caregiver in particular would attempt to chase us away. The ducklings on their part have faithfully kept close physical proximity to the caregivers, demonstrating clear signs of what has been termed imprinting.

Today is also a sad day. A mother sheep has been bleating for close to twenty-four hours on the nearby so called free-range farm. The farmer explained, when I went to inquire about the piercing bleating, that it was the ewe whose eight-week-old child was forcibly taken away from the mother or, as the farmer put it, the lamb had been weaned. This was one of the lambs I watched playing from my back deck the way people are used to seeing dogs play: chasing each other, mounting each other – the latter being more typical for male sheep – in the short-lived security of parental vicinity.

The wild duck population has since expanded substantially with many generations now residing here or regularly visiting the swimming pool turned duck pond, while the sheep-breeding facility has recently closed down. In a healthy environment nature and nurture work hand in hand. Building on biological dispositions, species-specific care provides the infant with the necessary kit for a favourable physical, psychological and social existence. For many animal infants, of both altricial and precocial species, adequate caregiving is not only crucial for their physical survival in the early period when they cannot fully fend for themselves in relation to the most basic needs such as protection and food, it also prepares the ground for successful coping as adults. In fact, as we will see, in this early period baby animals are dependent on the caregivers for the development of their own psychobiological regulation as well as the acquisition of more mundane skills such as those related to social and ecological competencies.

Weaning

Weaning of lambs and kids among the free-living populations may not occur until they are six months old. In later stages, as Cathy Dwyer points out, the suckling serves comfort purposes rather than nutritive ones.[2] In captive settings the average forced-weaning time is between eight and twelve weeks of age, while for sheep, goats and cows exploited for the production of milk and dairy items the young may be removed from their mothers within a day or so after birth. The forced removal disrupts the learning and social development of the young, and due to the psychological bond between the mother and the infant, as in other species, the separation is stressful for both parties.[3] The resulting distress vocalisations, this 'audiovocal thread of attachment'[4] that links the infant and the caregiver, is an essential feature for many animal species, enabling us to have our needs heard and possibly met in the most vulnerable period of our lives, but also later. Unsurprisingly, domestication and the repeated forced weaning of nonhuman animals, particularly in the milk-production industry, has not led to eradication of the social bond, despite the inconvenience the latter represents for these businesses. This is not difficult to understand if we consider, for example, that oxytocin, the hormone that plays an important role in attachment relationships, also triggers milk ejection.[5]

---

2   Dwyer 2009, 172.
3   Dwyer 2009.
4   Panksepp 1998, 262.
5   Weary and Fraser 2009.

The emotional pain animals feel following forced weaning may translate as production loss in economic language; therefore different techniques are employed to minimise the distress at weaning, such as the two-stage method of weaning lambs. Following this method, the lambs are not separated from the mother in the traditional abrupt way, but for a transitional period they are prevented from suckling.[6] Another staged method of weaning calves is fenceline weaning, where, again, suckling is prevented but the calf and the mother are able to remain in limited physical contact through the fence. When the fenceline method is impractical, farmers are advised to move the cows far away so they cannot hear their calves' vocalisations.[7] Another method employed for calves involves the use of mouth blocks, which prevent the calves from suckling without being separated from the mothers. These blocks come in various shapes and materials: for instance, they could be in the form of a plastic flap, but they could also be equipped with spikes. When the baby approaches to suckle the mother's milk, the spikes hurt the mother and she will push the baby away or move away from the baby. The spiky mount is, psychologically, one of the most torturous devices invented by humans for the 'management' of nonhuman animals. Through no fault of their own, the tender bond between mother and infant is violently broken as the infant suddenly becomes a source of physical pain for the mother, and the rejective mother a source of psychological pain for the baby, who will carry the scars into adulthood (if left to live past infancy).

As the literature and praxis around forced weaning confirms, animals' long history before domestication makes change difficult and slow:

We have controlled animals through domestication for only a few thousand years, and kept them in close confinement for only a few decades. The influence of the many millennia before domestication will heavily outweigh changes imposed during the last few decades.[8]

---

6   Schichowski, Moors and Gauly 2008.
7   Mathis and Carter 2008.
8   Broom and Johnson 1993, 33.

Animals' regulatory systems, including those of humans, which develop in infancy but remain open to change later on, are essential for animals' capacity to cope with the anxiety of novel situations, adverse situations, and stress generally. These capacities are heavily influenced by the developmental context. They also have phylogeny-based limits of plasticity: the adaptive threshold is much further out of reach than the animal-exploitative industries would find it comfortable and financially advantageous for the public to know when the economically biased euphemistic term 'animal welfare' is brought to the table. For instance, an egg-laying hen in a 'battery cage' remains a bird, and she *knows* that. Both pigs and hens, for example, feel the need to build a nest before giving birth or laying eggs even though, outside the reality of wilderness, where these needs evolved, they are not necessary.[9] A lion trapped in a circus or zoo enclosure remains a lion, and he *knows* that.

The justification continuously used by these institutions that the animals have 'not known' a different life hence there are no welfare issues with captivity is invalid. While it may be true that the animals have not ontogenetically known a different life, the phylogenetic knowledge – their biology and neurobiology – is equally relevant. Ontogenetic knowledge is short-term knowledge, acquired during life and mediated by changes in the nervous and immune systems, while phylogenetic knowledge is long-term knowledge, acquired over generations and mediated by changes in the DNA.[10] The latter has not changed significantly enough to enable animals a trauma-free life in captive situations that systematically deny such evolutionary knowledge. 'Some changes have occurred and others will continue to occur, but most characteristics are very resistant to change.'[11]

Social and ecological self-determination

Extreme physical constraints are but one of the pressing issues in animal exploitation. Other questions worthy of equal concern relate to the animals' need for agency and self-determination. This includes the

---

9   Broom and Fraser 2007, 15.
10   Århem and Liljenström 2008, 4–5.
11   Broom and Johnson 1993, 33.

capacity to control and be actively engaged in filial–parental, and other, social interactions and bonds. Keeping in mind the relatively short time nonhuman animals have been made conditionally or unconditionally subject to humans, it is unrealistic to expect that traits of agency and self-determination, which had in the animals' previous, captivity-free lives played crucial roles in the animals' survival and both their physical and social wellbeing, would suddenly vanish. Just as a human slave born, metaphorically speaking, with chains around her ankles knows the chains should be broken and she set free, despite her contemporary society teaching her otherwise, the same phylogenetic knowledge holds for other animals.

One of the most obvious indications of the need and indeed capacity for agency and self-determination is the seeming necessity to keep the animals tethered, encaged or otherwise restricted (even if only by fences). The absence of this might result in the animals making a different choice, for example to escape. Other indications that these traits have not left the captive animals' physical and mental domains are provided, among others, through food choices that they make when such a choice is available to them, and through the selective social-bonding systems that animals practise, once again, when such an opportunity is given to them in captivity. Both are critically dependent on the early developmental context, involve cognitive and emotional processing, and determine the animals' physical and mental wellbeing.

Food choices

In a healthy and natural environment, young animals are heavily dependent upon the caregiver(s) to provide them with initial cues in regard to food acquisition, and the access to a wide variety of food options is of critical importance. Lambs, for example, tend to eat the foods they see their mothers eat, and develop long-term preferences based on this early life experience.[12] While older animals' dietary habits tend to be less flexible than those of young animals, generally animals in the wild, and in captivity when such opportunity is offered, would exercise their choices based mainly on the nutritional value of the

---

12   Nolte, Provenza and Balph 1990.

foods, and on the interaction between palatability and post-ingestive feedback; that is, favourable or adverse reaction of the body to different foods experienced in the past.[13] For novel foods, while a certain amount of neophobia (the fear of unfamiliar food) is typical as well as useful to avoid potentially dangerous, poisonous foods, a phenomenon known as stimulus generalisation helps animals to infer from past experience and recognise positive or negative cues in the novel food. This motivates the animal's consumption or avoidance of the novelties.[14] The developmental context, however, can impair this phylogenetic bias for healthy, nutritional and species-normative food, testifying once again to the power of the early attachment and formative experience. Such was the sad case of Billy, for example, a chimpanzee who was reared in a human context and who, along with other irreparable scars captivity left him with, would prefer unhealthy human 'junk' food to the overall more appropriate chimpanzee-normative food varieties.[15]

The undeniable existence of favoured foods is often used in positive-reinforcement training and in experiments. For instance, based on preferred foods, a study of brown capuchin monkeys revealed that the monkeys respond negatively to unequal treatment.[16] Namely, the monkeys who were rewarded with cucumber for the same effort, or a bigger effort, that secured another monkey a grape – grapes being generally more favoured than cucumber – subsequently refused to participate, and even threw the cucumber back at the experimenter as an act of protest.[17]

While such research methods and captivity itself are ethically questionable, as the monkeys themselves indicated, food preferences and the related choices animals, including humans, make in life are of

---

13 Nolte, Provenza and Balph 1990; Provenza 1995; Cox et al. 1996; Bacon and Burghardt 1983.
14 Catanese et al. 2012.
15 For a detailed account and psychological analysis of Billy, who was reared in an intimate human context and lived there for fifteen years, then sent to an experimental laboratory, where he spent the next fourteen years to eventually end up in a sanctuary, see Bradshaw et al. 2009.
16 Brosnan and de Waal 2003.
17 A video excerpt of the experiment is available on the following website, from minute thirteen on: https://bit.ly/2GGkQU9.

vital importance. Exposure to a wide range of foods, which an animal's natural, non-captive environment would normally be able to provide, and the social and/or parenting group encourage, increases animals' feeding-behaviour-related cognitive capacities and decreases neophobia, making them fitter and more adaptable. In captive settings, exposure to food variety is limited, and sometimes monoculture is the only option. The possibility of developing sensory-specific satiety[18] aside, such non-exposure and the resulting stronger neophobia can be particularly stressful when anthropogenic circumstances force animals into a situation of sudden need for feed-related adaptation. Such could be the case following forced weaning or transfer to a feedlot, where the food might be unfamiliar to the extent of not enabling the animals any kind of inference from past experience, leaving them starving until they are able to familiarise themselves with these novel feeds.[19]

Sociality

Along with food preferences, group-living animals also exhibit social preferences. These preferences are based on familiarity, size, personality attributes, family relationships and other relevant characteristics.[20] An abundance of stories of inter- and intraspecific pair and group bonding comes from animal sanctuaries. Holland gathered some of these stories in her book *Unlikely Friendships*. Not uncommon are accounts of animals refusing to mate with the partners humans have designated for them in captive situations, such as the case of a female cheetah reported by Kaufman in the *New York Times*. The same article informs the reader that African penguins, who are generally monogamous, are particularly difficult to please, with an approximate twenty-five per cent refusal rate. Animals such as cows, turkeys and others in commercial settings are subjected to systematic rape, known under the technical term of artificial insemination, but for many exotic animals humans lack the knowledge to execute this procedure successfully, the article notes.

---

18   Animals who consume the same kind of food for long periods can become
      tired of that food and desire something new (neophyllic reaction).
19   Catanese et al. 2012.
20   Bode, Wood and Franks 2011.

It is becoming increasingly evident that nature is not as red in tooth and claw as has long been held. Instead it is cooperation, aided by social bonds, that pervades and determines success. As Broom summarises,[21] different nonhuman animal societies – much like different human societies, but not all – look after their weak and elderly. Besides the probable existence of emotional bonds, an old and weak individual in a group might be valuable in a time of shortage as they might remember some source of food or water, which would otherwise be difficult to locate. Several animal species also help rear infants who are not their own: for example, elephants, chimpanzees and other primates, and different bird species, including the kookaburra. The painted hunting dogs also have a similar arrangement, with pack members bringing food back to the pups and to the adults who look after the pups. Group-living prey species may strive to keep their weak members – for example, infants – inside the group, not on the outskirts where they could become easy meals for predators. When a bunch of strong-looking bison stand up to hungry wolves or even charge, it may be wise for the wolves to try their luck elsewhere.

Many examples of cooperation and friendships are included in Anne Innis Dagg's book *Animal Friendships*. The book is a telling collection of stories of animal friendships (based on scientific observations of animals in their natural habitats), which go far beyond the mere purpose of physical survival and reproductive success – the classical view of animals' actions and interactions. Among others, Innis Dagg tells of Laysan albatross pairs, who live for sixty-five to seventy years, sharing their lives and parenting duties. Amusingly, Innis Dagg also recounts that during her time as a first lady, Laura Bush, wife of former US president George W. Bush, praised these couples' commitment, which she felt could serve as a good example for humans. What Bush did not know, and it has only become known recently with DNA analysis, is that close to a third of these couples consist not of a male and a female, but of two females.[22]

The more we learn about the potency and critical functions of intersubjective attachment for both developing and mature organisms

---

21 Broom 2003, particularly 45, 53, 63–7.
22 Innis Dagg 2011, 68.

the more we understand the grief that follows the discontinuation of such relations among animals, whether from anthropogenic or other causes. This chapter discusses relevant aspects of attachment relations in some detail in an attempt to show that while outer expression of grief may vary across species and individuals, the feeling itself is comparable.

The emergence of attachment theory

Behaviourists and Freud

Not long ago behavioural scientists held the view that social bonds were a result of positive reinforcement: conventional rewards, such as food, induce the infant to become attached to the provider of such rewards – the caregiver. The infant's instinctual response to being fed and the pleasure derived from it become associated with the caregiver and consequently the caregiver herself or himself becomes a pleasant experience. Mary Ainsworth and Cara Flanagan provide helpful overviews of these theories.[23] Two views of the social learning theory of dependency – 'dependency' being the term for attachment at the time – were prevailing: one group viewed attachment as a label applied to certain learned behaviours, while the other group considered attachment an acquired or secondary drive.[24]

According to the 'secondary drive hypothesis', as Ainsworth summarises,[25] an infant has primary drives arising from basic physiological needs, such as the need for food when hungry. For the gratification of these motivational states/drives, the infant is completely dependent on the caregiver. Through repeated satisfaction of the infant's primary drives provided by the mother, the mother as the gratifying agent becomes a secondary or learned drive. Her presence is associated with forthcoming gratification, the infant is thus motivated to seek her presence, and becomes distressed if separation occurs. Through learning, new behaviours are added – including seeking

23 Ainsworth 1969; Flanagan 1999.
24 Reported in Ainsworth 1969, 982.
25 Ainsworth 1969, 982–3.

attention, help and approval – and generalised to encompass, in addition to the mother, other members of proximal society, such as fathers, teachers, siblings and other people. This view of dependency and its origins developed mainly from Clark Hull's behaviour theory, and was heavily influenced by Freudian theory.

Freud bases attachment on the infant's instinctual need for oral gratification. The person who satisfies this need in the first, oral stage of the infant's development becomes a love object and forms the basis of all further attachments in life. Accordingly it was believed that unhealthy attachments in one's life could result from deprivation or overindulgence in the oral stage.[26]

While it would be hard to deny the importance of nourishment, research has confirmed the biblical maxim that man cannot live by bread alone,[27] or 'by milk alone', as psychologist Harry Harlow put it.[28] Behaviourists themselves did not suggest that feeding was the sole basis of attachment, yet the findings from René Spitz's research concerning institutionalised human children in the 1940s, from Harlow's cruel experiments on rhesus monkeys and the existence of 'imprinting', as well as findings from Rudolph Schaffer and Peggy Emerson's study on attachment and food acquisition,[29] significantly undermined the importance of oral and nutritive-drives gratification and the projected consequences.

Early social deprivation

Spitz's studies in child development demonstrated that food and water are no guarantee for the child's wellbeing and indeed survival. Despite high standards of physical care, the lack of social bonds can result in depression, illness and even death. Spitz presents a few case studies

---

26   Flanagan 1999, 98–9.
27   Deut. 8:3.
28   Harlow 1958, 677.
29   Schaffer and Emerson 1964. Schaffer and Emerson, exposed to Harlow's research, and willing to further Bowlby's proposition of innateness of social tendencies, conducted a longitudinal study on sixty infants over a period of two years in Glasgow. The study confirmed that there was no direct link between attachment formation and oral gratification; that is, infants became attached to people who did not participate in feeding and other forms of routine care.

of deprived children in the 1952 film *Psychogenic Disease in Infancy*.[30] Especially disturbing are images of infants regressing emotionally and cognitively due to the lack of social bonds and stimuli, with death as a not-infrequent result. Floyd Crandall already in an 1897 editorial refers to 'hospitalism' as a 'disease more deadly than pneumonia and diphtheria.'[31] Hospitalism describes the physical deterioration in children following prolonged hospitalisation, which begins with atrophy unrelated to organic disease and can end in death if the child is not immediately returned to the familial environment. This of course had, and still has, important implications for the rearing of infants generally, though it has been particularly relevant in eras and/or places when children's institutionalisation in the form of orphanages constituted widespread reality. The infamous situation of the orphans in Romania, revealed after the fall of Ceausescu's dictatorship in 1989,[32] has offered research opportunity on effects of social deprivation aplenty. This includes a more recent study by Sheridan and colleagues on differences in neural development as a result of exposure to institutionalised conditions.[33]

In regard to nonhuman animals, Harry Harlow's experiments on monkeys showed no relation between attachment and feeding. As the author-perpetrator put it:

> We were not surprised to discover that contact comfort was an important basic affectional or love variable, but we did not expect it to overshadow so completely the variable of nursing; indeed, the disparity is so great as to suggest that the primary function of nursing as an affectional variable is that of insuring frequent and intimate body contact of the infant with the mother.[34]

---

30 The film is available online: https://bit.ly/33GFH2w.
31 Crandall 1897.
32 An undercover investigation revealed that the situation was still ongoing several years ago (Rogers 2009); however, there are serious attempts to close down all such institutions (Ironside 2011).
33 Sheridan et al. 2012.
34 Harlow 1958, 677.

Harlow's ethically unsustainable methods involved removing baby rhesus monkeys from their natural mothers and exposing them to mother surrogates. Two types of mother surrogates were developed: an unpleasantly hard surrogate made of wire-mesh, and a soft, cloth surrogate able to supply higher contact comfort than the wire version. The infants were divided into different study groups. The surprising result mentioned above was the absence of attachment to the wire figure even when 'she' was the sole food provider. The infants would eat from her, but when the infants, for example, were presented with fear-inducing stimuli, they would seek safety and comfort with the cloth mother surrogate regardless of the nursing situation. Furthermore, with time the infants' responsiveness to the lactating wire surrogate diminished, and their responsiveness to the non-lactating cloth mother increased. This further questioned the idea of attachment as a secondary drive following food conditioning.

Harlow continued his experiments on social deprivation and isolation in monkeys, in full knowledge that they were cruel.[35] This research showed the debilitating effects of social privation on infants.[36] Even though this research (and other research that followed) was motivated by the desire to understand the phenomenon and possible reversibility to help humans reared in isolation or other socially deprived settings, such as orphanages, it has important implications for all human-dependent animals, including orphaned wildlife who continue to be raised with minimal human contact, presumably to preserve their 'wildness'.

---

35 Psychologist Lauren Slater (2004, n.p.) reports that at a conference, Harlow responded to a colleague who reminded Harlow that he should be using the term 'proximity' instead of 'love', as follows: 'It may be that proximity is all you know of love – I thank God I have not been so deprived.' Slater, citing Rosenblum, also points out Harlow's refusal to use euphemistic terms in his research; for example, instead of 'terminate' he would use 'kill'; instead of 'restraining device' to refer to a device he used to force female monkeys to mate, he would use the term 'rape rack'.

36 Harlow and Harlow 1962; Harlow, Dodsworth and Harlow 1965.

Human children reared in isolation

In the 1940s two cases of human children reared in isolation were discovered, Anna and Isabelle, both aged six. As Davis reports,[37] Anna was born as an illegitimate child. Banned because of this from the grandfather's property where the mother was living and working, Anna spent her first few months in different institutions, before being handed over to a couple who neglected her. At the age of five and a half months she was returned to the mother and grandfather, who kept her in a room, providing only just enough care to keep her alive. When Anna was found, nearly six years later, she could not walk, she lacked speech, and was severely undernourished with bony legs and a bloated abdomen. Anna made substantial progress in the following years but did not achieve 'normality' by the time she died, aged approximately ten and a half years. It remains unclear whether she was originally intellectually unprivileged or whether such result was entirely due to environmental factors.

Isabelle[38] was also the product of an illegitimate encounter and spent most of her life in a dark room, but, unlike Anna, Isabelle shared the room with her deaf-mute mother. When Isabelle was discovered, she could not speak and showed no signs of hearing capacities, she could not walk and she had severe rachitis due to poor diet and lack of sunshine. Initially, specialists believed her 'feebleminded', but a systematic and apt (re)habilitation program, which Anna did not receive, soon started to produce results. Within two years, Isabelle reached normal mentality. The other advantage Isabelle had over Anna, as indicated above, was the presence of the mother. Though far from adequate, and even though Isabelle showed no awareness of relationships when tested soon after her rescue, such presence should not be ignored as some form of attachment likely developed between the two. Similarly, in the case of male twins discovered in the 1960s after spending their first eighteen months in an institution and then the following five and a half years reared in isolation, the fact that they had had each other offered the possibility of attachment.[39] When they were

---

37 Davis 1947.
38 Davis 1947.

discovered in 1967, the twins could barely walk, suffered from rachitis, were afraid of people, and communicated with each other mainly in gestures. They were first placed in a children's home, where they made good progress, to be later adopted into a household of two women, where they enjoyed a stable and loving environment, and were able to flourish both cognitively and emotionally.

Such accounts of successful outcomes should, however, be taken with a certain amount of caution. Physical and intellectual improvement and seemingly good social adaptation do not necessarily exclude the resurfacing of trauma later in life, or the presence of more or less disabling emotional states and choices, which may be hidden from the public eye, but often result from certain types of attachment styles. Also, in a sense, the actual damage can never – or certainly not without significant difficulty – be established, since stress and trauma can transmit across generations.

What the above cases demonstrate, however, is how important the presence of attachment relationships is, and how much effort is needed to achieve some kind of reversal when severe (de)privation occurs. In both Isabelle's and the twins' situations – but not in Anna's situation– the potential to form attachment was present since the isolation was not complete. Anna Freud and Sophie Dann conducted a study on six children who during their early years spent in a concentration camp only had each other, and found that despite severe deprivation they recovered well, presumably due to peer relations.[40] Similarly, Harlow and Harlow found peer support beneficial in reducing symptoms of (de)privation in monkeys.[41]

John Bowlby and attachment theory

As Ainsworth describes,[42] when referring to infant–mother ties and the implications thereof, three terms were commonly used, congruent with relative theories: *object relations* in the psychoanalytic view;

---

39   Koluchova 1972.
40   Freud and Dann 1951 (reported in Flanagan 1999, 58).
41   Harlow and Harlow 1962.
42   Ainsworth 1969, 970.

dependency in social learning theories;[43] and *attachment* proposed by Bowlby[44] when he was advancing his ethological approach to the study of the origin of infant–caregiver relationships, as a term devoid of theoretical connotations existent around 'dependency'. The current use of the term 'attachment' in psychology – even though it had occasionally been used before – springs from Bowlby. The term was first adopted by some ethologists, followed by psychologists studying animal behaviour and then spread to the rest of the field.[45] Nonetheless, later psychologists began to avoid applying the term 'attachment' to nonhuman animals, with the exception of primates, and the term 'social bonding' became preferred.[46]

Flanagan sees a distinction between attachment and bonding based on the cognitive processing involved: bonding being less of a mental process and representing the initial stages of the formation of a relationship, which includes the early period of infant (human inclusive) development, with attachment being the end of this process.[47] A not entirely dissimilar view is held by Ainsworth:

'Attachment' refers to an affectional tie that one person (or animal) forms to another specific individual. Attachment is thus discriminating and specific. Like 'object relations', attachments occur at all ages and do not necessarily imply immaturity or helplessness. [...] 'Attachment' is not a term to be applied to any transient relation or to a purely situational dependency transaction.[48]

Due to the nature of the social bonds that permit the experience of grief, the term 'attachment' is used in this text for such bonds for both human and nonhuman animals.

---

43 Dependency is not to be confused with the state of dependence, which implies helplessness and where the focus is the needed help, not necessarily the person who gives it, even though the latter does in part represent the early stage of dependency relations with the caregiver.
44 Bowlby 1958.
45 Ainsworth 1969, 971.
46 Flanagan 1999, 25.
47 Flanagan 1999, 25.
48 Ainsworth 1969, 971.

Mary Ainsworth worked closely with John Bowlby on the development of attachment theory and research; their partnership spread over forty years and several continents. Through her field work, Ainsworth contributed valuable empirical support to Bowlby's theoretical work.[49] While their approaches and conclusions have been complemented, adapted and even criticised, their contribution is significant, both as an incentive to a vibrant body of research in the area that has sprung up over the past decades, as well as for their observation of a more cross-species paradigm in relation to internal states.

Unsatisfied with the psychoanalytic and behaviourist theories of the time on the origins of attachment, and with their inability to explain the empirical data demonstrating adverse effects on the development of infants subjected to separation from caregivers, Bowlby's breakthrough is owed to Konrad Lorenz's and other ethological inquiries.[50] Lorenz was an Austrian ethologist, arguably most notorious for his work on imprinting in geese and other precocial species.[51] This work fascinated Bowlby, who saw striking behavioural similarities between imprinted chicks and attached human infants in relation to proximity-seeking and separation distress, and he was also intrigued by the evidence indicating that food provision was not essential for bond formation.[52]

Drawing on psychoanalytic and ethological principles, Bowlby developed a theory of attachment,[53] which posited that tendencies to form social bonds were adaptive and innate, with both caregiver and infant being predisposed for such a relationship. Infants possess a repertoire of innate behaviours (for example, crying) that elicit the caregiver's responsiveness. In evolutionary terms, the infant's potential for survival increases by obtaining adequate care, which in return benefits the survival of the gene pool and species.[54] One such example is the sea otter, whose infant's vocalisation enables the mother to find the young after resurfacing in the open sea following her search for

---

49   Ainsworth and Bowlby 1991.
50   Ainsworth and Bowlby 1991; Bretherton 1992.
51   Lorenz 1937.
52   Ainsworth and Bowlby 1991.
53   Bowlby 1969; 1973; 1980.
54   Flanagan 1999, 100, 102.

food; without such, the infant could be forever lost.[55] As Ainsworth and Bowlby elucidate, the traditional view saw dependence as an unavoidable feature in infancy, no biological value was attributed to it, and over time dependence was supposed to recede to give way to independence.[56] Bowlby, on the other hand, considered attachment as a major behavioural component, compared to feeding and reproductive behaviours, its principal biological function being that of protection, not only in infancy but also in adulthood. However, realising the multiple benefits of attachment relations and how they all contribute to evolutionary advantage, he later abandoned the notion of protection as a *principal* function.[57] Bowlby also observed the connection between attachment and exploratory behaviour and how the child's internal organisation develops within the infant–caregiver relation and interaction, as opposed to being the result of infants' projections of their own states as some postulated at the time, treating an individual 'as though he were a closed system little influenced by his environment'.[58]

Even though Bowlby's theory was inspired by ethology and rooted in an evolutionary paradigm, and even though attachment-related experiments on nonhuman animals continued, and still do,[59] for a long time attachment, and psychology more broadly, remained largely excluded from ethological discourse due to the ban on discussions of nonhuman animals' subjectivity. Yet behavioural evidence coupled with data from brain studies in other animals, including attachment-related processes, as well as the recognition of the primacy of subcortical regions and the implicit domain offer solid grounds for allowing inference in psychological matters from humans to other animals. The developmental context with its formative capacity for stress regulation

---

55 Panksepp 1998, 262.
56 Ainsworth and Bowlby 1991.
57 Reported in Cassidy 2016.
58 Bowlby 1973, 173.
59 Among the recent more controversial experiments that attracted public attention was a study that originally featured maternal deprivation in monkeys, approved in 2014 by the University of Wisconsin Madison animal-research committee. The maternal-deprivation part was subsequently dropped from the study, for scientific reasons not due to public pressure, as explained by lead researcher Dr Ned Kalin (reported in Wahlberg 2015).

or dysregulation critically affects resilience or vulnerability of animal minds, human and others. It is within this framework that Bradshaw articulated the trans-species approach and began exploring psychopathologies in elephants first, and then, with colleagues, in other species.[60] This framework also informs cross-species grief studies. Clearly, attachment relations are a prerequisite for the experience of grief, and, as we will see later, the developmental context, along with internal working models that emerge within it, may influence animals' responses to grief as pathological or not.

Attachment and psychobiological regulation

As cross-species research shows, infant animals are not only dependent on the caregiver for basic physiological needs, such as food and physical protection, the caregiver's input is also critical for the infant's maturing brain and self-regulation with long-term effects on the individual's physical and mental health. The brain and the patterns it forms are experience-dependent;[61] that is why these early interactions are so important. The caregiver is a source of stimuli to which the infant responds as well as a superstructure that informs the infant's organisation of such responses.[62] The caregiver effectively serves as an external regulator of the infant's biobehavioural modulation, which plays a critical role for the ripening emotion-processing circuits and the development of affective and neuro-endocrinological self-regulation.[63] The caregiver acts, in conjunction with the infant's own system, as a regulatory power for essential developmental processes, which affect both the individual's psychobiological adaptiveness as well as gene expression.[64] '[T]he adult's

---

60 e.g. Bradshaw 2005; Bradshaw 2009; Bradshaw et al. 2008; Bradshaw, Yenkosky and McCarthy 2009.
61 Differences in pre- and post-natal brain growth of course exist between precocial and altricial species with the first ones exhibiting higher levels of brain growth in the pre-natal stage and lower levels in the post-natal stage, while the opposite holds for altricial species (Bennett and Harvey 1985).
62 Polan and Hofer 2016, 120.
63 Schore 2005; Bradshaw and Schore 2007.
64 Bradshaw and Schore 2007.

and infant's individual homeostatic systems are linked in a superordinate organization,' explains Schore, 'that allows for mutual regulation of vital endocrine, autonomic, and central nervous systems of both mother and infant by elements of their interaction with each other.'[65] As already noted, these patterns of action and reaction become deeply ingrained, informing the individual's modus operandi intra-organismically (for example, levels of anxiety) and in relation to the outside world (for example, social competencies). Having been established before the individual's capacity for autobiographical memories, they are also hard to access and modify later in life should the necessity arise.

An attuned caregiver ensures a synchronised interaction, which maximises positive arousal states of the infant and mediates negative states.[66] In rats, for example – rats being a popular species for invasive attachment-related research – the infants' homeostatic system remains open to the mother's bioregulatory influence even after the infants have begun to consume solid foods and are capable of surviving on their own.[67] For constructive regulation, caregivers must themselves be able to regulate their own affective arousals. Affect regulation following a negative arousal state 'teaches' the infant that repair is possible; that is, that negative states and relational stress can be tolerated and regulated, leading, in the long run, to adaptive physical and mental health.[68] This so-called secure attachment promotes self-reliance in terms of capacity for affect regulation. Self-reliance, however, should not be understood as self-sufficiency; rather, through positive and attuned interactions with the caregiver, the maturing organism learns that support will be available in case of need. A self-reliant individual, Bowlby explains:

> proves to be by no means as independent as cultural stereotypes suppose. An essential ingredient is a capacity to rely trustingly on others when occasion demands and to know on whom it is appropriate to rely. A healthily self-reliant person is thus capable of exchanging roles when the situation changes: at one time he

---

65 Schore 2005, 207.
66 Schore 2005.
67 Hofer 1984, 186.
68 Schore 2005, 207.

is providing a secure base from which his companion(s) can operate; at another he is glad to rely on one or another of his companions to provide him with just such a base in return.[69]

Affect regulation in infancy thus cannot only be viewed as minimisation of negative emotional experience. Instead, affect regulation contributes to the development of an experiential matrix of an animal (human or other), which creates a sense of safety (or lack thereof) for the infant with lifelong implications. Adults with secure attachment histories are more likely to seek and find social partners capable of mediating their affective states. Social affect regulation, as studies suggest, appears to involve predominantly bottom-up processing, in contrast to self-regulation with its greater reliance on top-down processes.[70] Top-down processes describe cortical control of lower brain regions and physiology, whereas bottom-up refers to the opposite path. Coan suggests that this may make social regulation more efficient and less costly compared to the self-regulatory path involving suppression of emotion and cognitive reappraisal.[71] He also adds mindfulness meditation to the list of top-down affect regulation strategies; however, this position has been challenged, and I return to this in Chapter 4.

Poor and dysregulated transmissions within attachment relations, on the other hand, are equally 'affectively burnt in'[72] the infant's maturing brain. The brain consequentially develops internal working models based on insecurity with potentially vulnerable psycho-physiological outcomes, including pathological responses to death of proximal subjects.

The infant's primary aim in the initial period of life thus is the establishment of emotional communication with the caregiver, which will facilitate the developing self-regulatory systems,[73] for better or for worse. So strong is this innate need for connection that in these early stages of animals' lives avoidance learning is inhibited and preference

---

69   Bowlby 1973, 359–60.
70   Reported in Coan 2016, 258–9.
71   Coan 2016.
72   Bradshaw and Schore 2007, 429.
73   Schore 2005, 206.

learning is strongly dominant.[74] Since the amygdala, which will later play a critical role in detecting aversive stimuli, is not fully developed in neonates, it is possible, Coan notes, that at this stage all stimuli are only encoded as either familiar or non-familiar.[75] This would enable young animals to form attachments to both loving and abusive caregivers. The latter, however, not only provides fertile grounds for dysregulation and the development of psychopathologies but it also makes them vulnerable to abuse later in life.[76] As Polan and Hofer explain, summarising relevant research in rats but applicable also to other species: 'early maternal maltreatment is a double-edged sword: One edge induces adult depression, whereas the other provides a safety signal in the adult that blunts new fear learning and relieves the depressive behaviours.'[77] These early experiences constitute building blocks of internal working models and related attachment styles, to which we turn shortly. Before that let us have a taste of some of the things happening at the neural level.

We noted earlier that critical information about attachment mechanisms comes from invasive brain and other forms of research (for example, behavioural) on nonhuman animals. While there are species differences in the detailed expressions of such mechanisms, the principles are largely applicable across species. The main goal of such research is to understand and ultimately help humans but of course it also helps predict the experiential matrix of nonhumans. As Coan reminds us,[78] given the numerous neural structures involved in attachment relations, the entire brain can be viewed as an attachment system; however, we can single out a few structures and processes in an exemplary capacity.

The stress-response system, for instance, is heavily influenced by attachment relations. Comprised by the autonomic nervous system (ANS) and the hypothalamic-pituitary-adrenal (HPA) axis, hyper-activation and dysregulation of the system can lead to a number

---

74 Polan and Hofer 2016, 121.
75 Coan 2016, 249.
76 Polan and Hofer 2016, 121–3.
77 Polan and Hofer 2016, 123.
78 Coan 2016, 244.

of physical- and mental-health problems. The ANS is the part of the nervous system that supplies internal organs and regulates certain bodily processes, such as heart rate, blood pressure, digestion, etc. When the ANS receives information from the body and the external environment, it stimulates or inhibits bodily processes through its two main branches: the sympathetic branch (the so-called 'fight or flight' response) and the parasympathetic branch (the 'rest and digest' response). The developmental context may affect individual differences in the autonomic response to stress. An adverse environment (for example, physical or emotional abuse, prolonged separation from the caregiver, unresponsive caregiver) can sensitise the organism to stressors, making it hyper-reactive and maladapted to constructively deal with challenging circumstances. The sympathetic branch may become dominant, leading to hypervigilance, anxiety, aggression, sleep problems and other physical and psychological challenges.[79] Stress in the early period also affects animals' immunological resilience, making them more prone to inflammation later in life.[80] Attachment and social interactions more broadly continue to affect the ANS in adulthood. Positive social relations have a calming effect as they can reduce stress-related activity in the ANS and the HPA axis[81] with negative interactions achieving the opposite.

The HPA axis is a neuroendocrine system (combining the endocrine system and the nervous system) of fundamental importance for stress and behaviour regulation. When a threat is detected, the corticotrophin-releasing hormone is released from the hypothalamus. This induces the pituitary gland to release the adrenocorticotropic hormone (ACTH), which stimulates the adrenal cortex to secrete glucocorticoid hormones, such as cortisol or its functional parallel corticosterone in rodents, birds, reptiles and amphibians. These then circulate throughout the body and brain. Resources are thus mobilised for the organism to address the threat. For instance, cortisol raises blood sugar and adrenaline levels, and suppresses growth, digestion and other things that are not crucial in an emergency and that the

---

79   e.g. McLaughlin et al. 2014; Alkon et al. 2017; Gander and Buchheim 2015.
80   For a review see Ehrlich et al. 2016.
81   Coan 2016, 243.

organism can attend to once the emergency is over.[82] Critically, the HPA axis is directly influenced by the early social context. Repeated HPA activation and high corticosteroid levels, Bradshaw and Schore explain, can impair gene expression in neurogenesis (formation of new neurons) and synaptogenesis (formation of neural synapses) as well as influence the memory, cognition and affect regulation circuits in the post-natal maturing brain.[83]

The hypothalamus, apart from being involved in the production of cortisol, produces and, through the posterior pituitary, secretes a number of other hormones, including the social neuropeptides oxytocin and vasopressin. These peptides, and their non-mammalian correlates (vasotocin, and oxytocin-like isotocin and mesotocin), play an important role in social relations across all vertebrate species.[84] Oxytocin has gained some popular attention as the 'love hormone' since its levels increase during intimacy of various kinds; for example, sexual, nursing, platonic. Oxytocin is also involved in parturition (exogenous oxytocin is sometimes administered to aid labour) and in triggering lactation, as well as playing a significant role in affiliative behaviour, social preferences, pair bonding and maternal behaviour.[85] It also mediates fear responses of the amygdala.[86]

Oxytocin and vasopressin activity along with the release of endogenous opioids following a positive social experience consolidate associations of interactions with social partners as rewarding, enabling attachment relations across species to form and continue beyond the infant–mother dyad.[87] Coan points out that while attachment theory accentuates the regulatory dimension of attachment relations and proximity seeking, proximity seeking as a means of getting the reward is equally important.

Reward seeking, however, could also be a double-edged sword. If the attachment figure is not available, as in the case of the death of such

---

82   Morris and Fillenz 2003; Mitrovic n.d.; Coan 2016.
83   Bradshaw and Schore 2007, 428.
84   For a review see Insel 2010.
85   e.g. Ross et al. 2009; Insel 2010; Coan 2016.
86   Knobloch et al. 2012.
87   Coan 2016, 250–1.

a figure, reward seeking can exacerbate the pain and complicate grief. For instance, studies have shown that reminders of the deceased loved one through photographic and verbal material activate brain regions linked to physical and social pain processing (such as the anterior cingulate cortex, insula and periaqueductal gray) in both people suffering from non-complicated grief and those suffering from complicated grief. But it was only in people with complicated grief that the nucleus accumbens produced significant activation.[88] The nucleus accumbens is a reward centre that activates in animals, including humans, not during the *experiencing* of a reward, but in *wanting* the reward. O'Connor and colleagues write:

> The addiction-relevant aspect of this neural response[89] may help to explain why it is hard to resist engaging in pleasurable reveries about the deceased even though engaging in these reveries may prevent those with CG [complicated grief] from adjusting to the realities of the present. Many who suffer from addiction-like disorders experience them as afflictions; similarly we are not suggesting that reveries about the deceased are emotionally satisfying, but rather may serve as craving responses that may make adapting to the reality of the loss more difficult.[90]

As we will see later, animals with insecure attachment histories are more prone to developing complicated grief.

Internal working models and attachment styles

Central to attachment theory are so-called 'internal working models'; that is, mental representations of the self, the socio-natural environment and the self in relation to the latter. Bowlby, who discovered the concept of working models in the writings of biologist J.Z. Young, postulated that animals develop two working models, an environmental model and an organismic model, which help organise

---

88   O'Connor et al. 2008; O'Connor 2012.
89   Knutson et al. 2001.
90   O'Connor et al. 2008, 972.

relevant information about the self and the world, informing expectations, beliefs, 'rules'. To successfully navigate in an environment thus, individuals must have knowledge of the environment as well as of their own skills and potentialities. 'The more adequate the model,' Bowlby writes, 'the more accurate its predictions; and the more comprehensive the model the greater the number of situations in which its predictions apply.'[91]

The caregiver is a key feature of the infant's world, providing not only comestibles and protection but serving as a 'surrogate' psychobiological regulation system for the infant's developing organism. In this delicate period through the interactions with the caregiver, the infant animal develops working models of the caregiver. Through the caregiver's reactions to the infant (for example, responsive or not, attuned or not, etc.), the infant also develops a working model of the self, as Siegel puts it eloquently (and previously cited): 'we first know ourselves as reflected in the other'.[92] Caregivers' responses thus provide the infant with a plethora of critical information about the self, including whether the infant is 'worthy' of love or not, whether help is available in case of need or not, how to behave to elicit it, etc. The infant will organise this information in the working model of the self, which will accompany the individual throughout life. This early priming also teaches the infant about social relations more generally; the infant–caregiver attachment relation becomes a model for other social relations/attachments in the future. As Bowlby notes:

> Confidence that an attachment figure is, apart from being accessible, likely to be responsive can be seen to turn on at least two variables: (a) whether or not the attachment figure is judged to be the sort of person who in general responds to calls for support and protection; (b) whether or not the self is judged to be the sort of person towards whom anyone, and the attachment figure in particular, is likely to respond in a helpful way. Logically these variables are independent. In practice they are apt to be confounded. As a result, the model of the attachment figure and the

---

91   Bowlby 1982 [1969], 81.
92   Siegel 2011 [2010], 62.

model of the self are likely to develop so as to be complementary and mutually confirming.[93]

Bowlby then gives two examples, one of an unwanted child and one of a much loved child. The unwanted child will likely not only feel unwanted by the attachment figure but will feel 'unwantable' generally, by everyone. The much loved child on the other hand will likely grow up feeling confident that they are lovable not only to the parents but also the rest of the world. 'Though logically indefensible,' Bowlby concludes, 'these crude over-generalizations are none the less the rule. Once adopted, moreover, and woven into the fabric of the working models, they are apt henceforward never to be seriously questioned.'[94]

These early interactions and the caregiver's responsiveness determine the young animal's attachment style as secure or insecure, which affects the kind of internal working model of the caregiver the infant develops. The classification of security of attachment, the tool to investigate it, namely the so-called 'strange situation', as well as the concepts of secure base and safe haven were Ainsworth's contribution to attachment theory.[95]

Secure base/safe haven. The attachment figure serves as a secure base from which the infant explores and a safe haven to which the infant returns.[96] Stimuli input is essential for learning and cognitive development as such. A happy and secure attachment relationship facilitates the infant's exploration of the social and physical environment and the resulting learning. The process of exploring is, by definition, saturated with novelty and thus engenders a certain level of insecurity. When the infant becomes uneasy while exploring, the reassuring and soothing presence of an attachment figure is most welcome. At the same time the availability of such a figure as a secure base encourages infants to explore further. As a consequence, securely attached infants are more willing to explore than insecurely attached

---

93   Bowlby 1973, 204.
94   Bowlby 1973, 205.
95   e.g. Ainsworth 1967; Ainsworth 1969; Ainsworth and Bell 1970; Ainsworth, Bell and Stayton 1971; Ainsworth et al. 1978.
96   Ainsworth 1969, 1006; Ainsworth and Bowlby 1991.

ones, and this might lead to cognitively divergent results. Harlow's invasive experiments with monkeys support this proposition, as after the infants had adopted the surrogate mother as a security figure their exploration and play behaviour increased.[97] However, this was the only benefit of the cloth mother; in other respects these monkeys developed social aberrance typical for monkeys raised in bare wire cages with a gauze diaper pad as the only source of contact comfort.[98]

Strange situation. The 'strange situation' is a brief, laboratory-set evaluation of the infant–caregiver attachment, in which the attachment style is determined by observing the infant's behaviour when the mother leaves the room and upon her return. Infants show individual differences in their responses indicative of differences in attachment styles. Ainsworth detected three different styles: secure attachment, insecure-avoidant and insecure-resistant/ambivalent/anxious. A number of infants from the studies, however, remained unclassified or were deemed hard to classify in the 'strange situation'. This prompted Main and Solomon to examine the evidence, which then led them to add a fourth attachment category: insecure-disorganised/disoriented.[99]

Ainsworth maintained, contrary to the general view of the time and congruent with Bowlby's own position, that adequate responsiveness and physical contact does not spoil the child, but it promotes a secure attachment, which further down the track translates into independence.[100] Unresponsive caregivers or caregivers who are inconsistent or rejecting in their care for the infant are likely to mark the infant with some form of insecurity. Linda Graham presents the four styles in a concise way, which I will now summarise.[101]

When the caregiver is attuned to the infant – available, and responds adequately to the infant's needs and is capable of both minimising distress and maximising joy – the infant will likely develop *secure attachment*: the infant will feel loved and safe, will begin to trust his/her own capacities to elicit response and will feel confident seeking

---

97   Harlow 1958, 684.
98   Harlow and Harlow 1962, 141–2.
99   Main and Solomon 1986.
100 Ainsworth 1991.
101 Graham 2008.

connection to the attachment figure, and importantly, the infant will learn that regulation of negative arousal states is possible. This will likely lead to a secure and autonomous adult[102] who will be comfortable with relationships and emotions generally, and be better able to self-regulate or seek help if needed.

When the caregiver is distant, emotionally or physically unavailable, avoidant, dismissive, rejective, unable to regulate and/or effectively engage in the infant's positive or negative arousal states, the infant will likely develop an *insecure avoidant attachment*: the infant appears indifferent to the caregiver; expecting rejection rather than a soothing or positive response; in self-defence the infant appears numb and does not seek interaction with the caregiver. This will likely lead to avoidant, dismissive adults who have difficulty trusting others; they may be aggressive and appear overly self-reliant; they are not comfortable with intimacy and may devalue relationships.

When the caregiver is anxious, unpredictable in their response, sometimes loving, other times harsh, sometimes too engaged, other times distant, the infant will likely develop an *insecure-anxious/ambivalent attachment*: the infant will be ambivalent about the availability and reliability of the caregiver; infants are difficult to soothe; sometimes they are clingy, other times resistant, basically mirroring the anxiety and ambivalence of the caregiver. As adults they will likely be anxious and over-preoccupied with the availability of the attachment partner, harbour a deep fear of abandonment, chronically vigilant about separation, they can become obsessed with the partner and show extreme jealousy.

Lastly, if the caregiver is, permanently or temporarily, disorganised, fragmented, or even abusive and thus frightening to the infant, the infant may develop *insecure-disorganised attachment*: the infant will feel helpless, disorganised, chaotic or paralysed, and won't be able to soothe. This may

---

102 Currently, for adult attachment assessment, a dimensional conceptualisation of attachment styles is preferred (though not exclusive) in lieu of the categorical one (low avoidance/low anxiety = secure, low avoidance/ high anxiety = preoccupied, low anxiety/high avoidance = dismissing/avoidant, high anxiety/high avoidance = fearful/avoidant). For an analysis see, for example, Fraley et al. 2015.

lead to a disorganised adult who has difficulty functioning, is incapable of emotion regulation; there may be fragmentation and dissociation.

It is critical to note, however, that these attachment styles (including insecure forms) may, at least temporarily, be adaptive. For the infant's immediate context they may be adaptive in terms of the infant's necessity to negotiate and bring some coherence to the caregiver's responses, whether supportive or not.[103] For example, a clingy infant may annoy an avoidant caregiver risking abandonment, and in that early period of life some proximity is better than no proximity; 'to the child it feels literally vital to find a way of living with this mother'[104] or other primary caregiver, because that is all there is. Caregiving styles promoting insecurity may also be adaptive in the sense that a secure attachment would benefit an infant in a secure and predictable environment, but other styles may be more efficacious in a chaotic and dangerous environment that requires constant vigilance.[105] Simultaneously, caregivers are themselves affected by both the external environment as well as by their own internal landscape, both of which guide their responsiveness to the infants' needs. In an environment that is inherently insecure (for example, food scarcity, violence, etc.) the caregiver does not have the 'luxury' to attend to the infant the way they may in a peaceful and plentiful context. Whether that context changes or not, the caregiving style and consequent attachment styles developed by the infant will become deeply ingrained in the infant and mark the infant for life. Invasive research in nonhuman animals, while ethically unsustainable, testifies to the paramount influence of the caregiving style on animal infants, which also transmits across generations.

Monkeys and rats continue to be favourite victims of attachment research. Most of it is conducted to extrapolate findings to humans with little empathy for the nonhuman subjects themselves and little consideration for all other animals, trapped in exploitative settings that range from depriving to outright torturous (farms, circuses, and others), to whom these psychobiological principles and processes also apply. Already Harlow and Harlow observed that emotionally deprived

---

103 Ehrlich et al. 2016.
104 Parkes 2009 [2006], 103.
105 Polan and Hofer 2016, 118.

monkey mothers ended up being either negligent or abusive to their own children.[106] Similarly, the early social context influences rats' lives and parenting styles: rat pups of mothers practising moderate (low) licking and grooming (LG) and low arched-back nursing (ABN), when placed in foster care of mothers who practised increased (high) LG-ABN, resembled biological pups of high LG-ABN mothers and vice versa.[107] Environmentally induced transmissions of individual differences in behaviour and stress responses can span across generations as the mothering style of the pups-turned-adults also resembled the style of the fosterer/caregiver, not that of the biological mother. That is to say, pups reared by high LG-ABN females as adults became themselves high LG-ABN mothers regardless of their biological origins. High LG-ABN is more beneficial for long-term stress and anxiety responses of rats (that is, more modest HPA response to stress compared to offspring of low LG-ABN mothers), their level of fearfulness, exploratory activities and cognitive development.[108] Other studies showed similar results. For instance, behavioural biologist Dario Maestripieri[109] took female infant rhesus monkeys whose biological mothers had a history of being abusive and put them in the care of non-abusive foster mothers, and, vice versa, infants from non-abusive biological mothers were reared by abusive foster mothers. About half of the infants of non-abusive biological mothers reared by abusive foster mothers became themselves abusive towards their own children, while *none* of the infants raised by non-abusive foster mothers ended up being abusive to their own children. Such developments, however, are potentially reversible. In the above study of rats the reversal was achieved through pharmacological intervention. *In natura*, while we might not be as preoccupied with the molecular and cellular aspects as the *in vivo* and *in vitro* investigations are, rescuers of animals (human inclusive) regularly face challenging physical, psychological and behavioural states of the rescued organisms, brought about by

---

106 Harlow and Harlow 1962.
107 Meaney 2001; Weaver et al. 2004; Sapolsky 2004.
108 Liu et al. 2000; Weaver et al. 2004; Sapolsky 2004.
109 Maestripieri 2005, reported in Suomi 2016.

privation, deprivation, neglect or direct abuse, which we strive to reverse. Sometimes we succeed in doing so, other times not.

Attachment and non-mammalian/non-avian species

While most of the research in the area of attachment and social bonds focuses on mammals and birds, and most humans believe these are the only taxa with capacities for intersubjective attachment, emerging evidence, both anecdotal and scientific, of affiliative relations among other vertebrates may in the future reveal a sophistication of affective bonds in those species that eludes our current understandings. People were endeared, and surprised, to learn about the strong bond the 130-year-old tortoise Mzee and the young hippopotamus Owen formed at the Kenyan Haller Park wildlife sanctuary in 2004.[110] Reptiles are not renowned for their gregariousness; however, much more could be hiding under the scaly coating than may at first appear to an uninformed observer. Turtles engage in social play and seek out specific individuals as playmates.[111] Some lizards live in family units consisting of a monogamous pair and their offspring.[112] Snake mothers can exhibit long-term maternal care and even look after other snakes' offspring, while adult male snakes may visit the rookery without any obvious interest in reproduction, and let the infants crawl and coil on and beside them without exhibiting antagonistic behaviour.[113] Interestingly, though not surprisingly, paleontological evidence suggests extended parental care in dinosaurs.[114] Crocodiles, too, may form attachments and mourn following a loss. When Buka, an elderly crocodile, died, Bonnie, Buka's partner of twenty years, was left in mourning, as reported by the ABC.[115] John Lever, who has been around crocodiles for forty years, has witnessed many instances of crocodile attachment (for instance, the crocodiles may sunbathe together with one resting their chin on the partner) as well as loss. When I spoke to him a

---

110 Ludden 2005.
111 Kramer and Burghardt 1998.
112 White, Uller and Wapstra 2009.
113 Amarello, Smith and Slone 2011.
114 Horner and Makela 1979; *BBC News* 2003.
115 Burt and Culliver 2020.

few days after Buka's death, Bonnie was hiding at the back and was not coming for food. Based on past experience, Lever predicted this behaviour would continue for three or four weeks.

Social bonds are not unknown in the world of fishes either. Some species (such as the maternal mouthbrooding tilapia), for instance, nurture an extended period of maternal care, with the infants bringing into the world strong predispositions concerning the mother. This facilitates bonding though appropriate experience/learning is necessary for these predispositions to be maintained and for the induction of (social) preferences.[116] Social bonds and pair bonds among fishes, especially in schools of fishes in the wild, are challenging given that studies of this kind are normally based on observation of physical proximity between individuals. With the aid of molecular analysis, however, it has been possible to determine that pair bonding does exist and that parental care, like in humans and several other animals, can be a shared duty between the mother and the father in fishes, too.[117] The importance of sociality in fishes is further accentuated by the negative side-effects isolation, particularly repeated periods of isolation, can have on fishes. Such is reflected in the story of the zebrafish who was discovered immobile at the bottom of the tank when conspecifics were absent but who resumed activity when other fishes were introduced to the tank. The zebrafish also reacted positively when human antidepressant drugs were added to the tank water. As a consequence of the psychobiological and behavioural comparability to humans, the zebrafish, in complete disregard of the fish's own wellbeing, becomes a new 'animal model' candidate for developing drugs for human psychiatric issues.[118] Nonhuman animals' psyches continue to be exploited for human interest but largely ignored in their own right.

---

116 Russock 1999.
117 Takahashi et al. 2012.
118 Brief summary of the findings reported in Cyranoski 2010; see also Abreu et al. 2018.

The picturesque fabric of loss

I explained earlier that I would be using Parkes's definition of grief, centred around separation distress; that is, an individual is grieving if he or she has experienced a loss, followed by 'a reaction of intense pining and yearning for the object lost (separation anxiety)'.[119] Previously in this chapter we have seen that attachment relations play a central role in the formation and functioning of animals' psychobiological matrixes. Myron Hofer believes that '[o]ne of John Bowlby's great contributions has been to place attachment and loss in the perspectives of development and evolution'.[120] For Hofer this means that he can study attachment and loss in invasive experiments on nonhuman animals and infer the findings to humans. Of course there is the obligatory warning to self and the reader that we must be cautious when generalising across animal species, followed by the fundamental recognition that 'evolution is conservative, tending to use what is available and adding new features in new species, but retaining much of the old'.[121] This recognition can be used to continue to subject nonhuman animals to cruel investigations into separation distress (and other subjective phenomena) or it can be used to learn to appreciate the potency of that effect on, and affect in, the animals themselves.

While humans' contemplations of their own mortality and similar philosophical deliberations can colour the loss in distinctive ways, much of the tragedy of loss manifests in the immediacy of the pain caused by separation distress. Attachment relations, both in infancy and adulthood, influence and sustain animals on multiple psycho-physiological levels, which are affected, separately and jointly, when loss occurs. The infant–caregiver dyad is of particular and vital relevance in terms of the infant's psychobiological and more broadly social development. The caregiver's presence does not only help to regulate the infant's physiology *generally*. The mother–infant interactions have been shown to affect a myriad of specific physiological mechanisms in animals ranging from heart rate to growth hormones

---

119 Parkes 2009 [2006], 30.
120 Hofer 1984, 183.
121 Hofer 1984, 188.

to sleep patterns.[122] The infant's system has the potential to collapse with the withdrawal of this multimodal support. Attempts at repair to promote recovery and healing would have to be able to address this regulatory variety.

Importantly, the homeostatic regulatory effects of various forms of input from the social environment are not limited to support for infantile animal organisms, but pertain to adult animals as well. Like the attunement between infant and caregiver, adult animals may also develop attunement and synchronisation between proximal subjects, mediated through various senses, such as smell, touch and vision. An optimal dynamic positively affects psychobiological regulation by providing either stimulation or modulation of arousal levels. This can prevent aversive under- or over-stimulation, keeping the individual in an optimal arousal zone.[123] Conversely, separation between attached individuals leaves the animal without a co-regulator, with effects on psycho-bio-behavioural outcomes. Further, Hofer also reminds the reader of the influence of social *zeitgebers* (literally, 'time-givers') in the regulation and synchronisation of internal rhythms.[124] Individuals living in social groups develop synchronised rhythms and when moved to a different group they adapt to the new group's rhythms. When social cues are withdrawn or heavily diminished, individuals' biological rhythms appear to desynchronise, with a potentially negative effect on homeostasis as well as on adaptive functioning. Since a companion may constitute a significant part of an animal's socioenvironmental fabric and contribute substantially to its dynamics, the sudden withdrawal of these cues, for instance following permanent separation due to death, could, as Hofer proposes, lead to desynchronisation with more or less severe negative outcomes. Outside the immediate framework of bereavement, relations between circadian rhythms and mood disorders have received some attention.[125] This is another facet of grief, which is

---

122 Hofer 2006.
123 Field 2011. Of course, adult attachment relations are also influenced by the individuals' attachment styles and working models, see, for example, Mikulincer and Shaver 2016.
124 Hofer 1984.
125 For a more recent review see, for example, Bechtel 2015.

unrelated to 'higher' cognitive capacities but might play a critical role in nonhuman animals' coping with loss in their social network, as do attachment styles, which we turn to now.

In December 2017 I was invited to submit a commentary to a recently published article on animal suicide.[126] I had never considered the possibility of suicide in nonhuman animals and was initially unsure what I could contribute to the discussion. Reading through the paper, in which the author, philosopher David Peña-Guzmán, argues convincingly for the possibility of nonhuman animal suicide, I realised that some of the suicidal behaviour could be the result of complicated grief. Preti, for instance, in his review of research on animal suicide between 1967 and 2007, mentions the seemingly puzzling phenomenon of dogs reportedly starving themselves to death following the loss of their human guardian.[127] I argue that companion animals and other captive animals are at higher risk of developing complicated grief (not all of which ends up with suicide of course). This proposition is based on evidence that individuals with insecure attachment histories are more vulnerable to complicated forms of grief, and on the fact that captive animals are not usually afforded a normative species-specific upbringing that may foster security of attachment. The insecure-anxious attachment style appears particularly problematic for grief because animals with this style are 'wired' for clinging. The brain becomes organised in such a way that it chronically searches for cues indicating the availability of the attachment figure, and quickly detects cues of unavailability.[128] With such brain organisation, when the loss is permanent (through death or other modes of separation) the cognitive recognition, the *knowledge*, of such may not be enough to override the underlying mechanisms of insecurity and the propensity/compulsion for searching. This may then lead to complicated grief and the inability to adjust to the loss – a condition sustained also by simultaneous occurrence of pain-related and reward-related neural activity, as mentioned earlier, producing a kind of craving for the object lost.[129]

---

126 Peña-Guzmán 2017.
127 Preti 2007.
128 Fraley and Shaver 2016, 44–5.
129 O'Connor et al. 2008.

If this obsessive cycle is not broken, it can severely impact on the individual's long-term psychological and physiological wellbeing.

Further, many rescued animals, both companion animals and others who live at sanctuaries, have not only had a suboptimal upbringing but have a truly traumatic past, including physical abuse. When they find some security with either a human or another animal, the loss may be particularly traumatic, especially if the rescued animal only becomes attached to one individual as opposed to forming a wider social circle with emotional investments in multiple individuals (sometimes, of course, the animal is not given a choice). The extent to which one's self and identity are merged with a single companion is, in fact, also predictive of grief responses, complications or lack thereof when this companion dies.[130] In the case of Chickweed and Violet (introduced in the opening paragraph of the first chapter), who grew up together and were inseparable until Violet's passing, it is safe to assume that their identities were highly merged, and Chickweed's grief response – his ongoing sadness and anger, which over time diminished but never truly left him – confirms it.

* * *

In this chapter we have become acquainted with mechanisms of attachment formation and loss from a cross-species perspective. The material presented highlights the vital importance of attachment relations for animals' biological and psychological development, and the consequences of attachment dysregulation and separation distress. Most of the detailed knowledge on the subject comes from invasive experiments on nonhuman animals. The BrainMind comparability across species allows researchers to infer from other animals to humans, while convention and speciesism continue to impede the full appreciation of nonhuman animals' experience of the joys and pains of attachment and loss as comparable to humans' own. However, there is little support for this prejudice in the neurobiological and behavioural evidence, which instead suggests that grief, when it occurs, is likely to differ on an individual rather than a species level. In the following

---

130 Fraley and Shaver 2016, 58.

chapter we explore some of the philosophical questions that emerge in considerations of nonhuman animal grief and look at mourning practices from a human cross-cultural perspective, a perspective that is often ignored when discussing nonhuman animal grief.

3
Cross-cultural grief matters

We bereaved are not alone.
We belong to the largest company in all the
world–
the company of those who have known
suffering.
— Helen Keller, 'We Bereaved'[1]

Nonhuman animals of diverse species have been observed exhibiting affective behaviour at the death of proximal subjects. Descriptions of nonhuman animals seeming to perform rituals, including burials, at the death of both conspecifics and members of other species, are also not unheard of. Neurobiological and ethological evidence aside, scientists and other commentators remain cautious in ascribing a human-comparable grief dimension to nonhuman animals. Such cautiousness is based mainly on the uncertainty of nonhuman animals' capacities for cognitive appraisal concerning death matters, including the question of nonhuman animals' awareness of death and their own mortality. It also appears to spring from the inaccurate position that there is universal consent within the human species itself on the substance of such awareness.

---

1   Keller 1929, 1.

So far we have reviewed and discussed the scientific evidence, specifically the neuropsychological foundations that predicate a human-comparable capacity for grief and feelings of profound loss in nonhuman animals. We have also addressed psychosocial mechanisms and the historical background that permit the disjunct between scientific evidence and public and scientific perception of nonhuman animal emotional capacities. Following a very brief overview of ethological observations of grief expressions, in this chapter we explore the issue of 'understanding' death, and then we journey to a selection of human cultures around the world whose responses to death and grief differ substantially from those of what could loosely be described as the modern-day Western secular human (of course, on this journey we will bear in mind that many such observations have been made by the WEIRD, including the present author). The latter – Western-like approaches – are often used in a comparative capacity to evaluate grief in nonhuman animals. If the same flawed criteria were to be applied in considerations of grief and its expression across human cultures, and supposing that we were unable to ask individuals from other human cultures about their feelings and for explanations of their rituals, restricting us thus to rely on behavioural observations only, it may leave us equally puzzled in regard to the presence and nature of their grief and awareness of death in those other cultures, as it does for nonhuman animals.

Ethological observations of nonhuman animal grief and loss

Elephants are one of the few nonhuman animal species that have been observed engaging in what can be taken to be explicit death rituals, which include covering the deceased body with branches and dirt, and lifting and manipulating the bones of relatives.[2] Magpies have been observed bringing grass and laying it beside a dead conspecific, then standing as if in vigil for a few seconds before flying off.[3] Similar scenes have been witnessed with ravens and crows recently[4] and possibly less

---

2   Reported in Bradshaw 2009.
3   Bekoff 2009.
4   e.g. Derbyshire 2009.

recently given their inclusion in religious narratives. For instance, according to an account in the Qur'an, God sent a raven to show Cain how to bury his brother.[5] A similar but more explicit account of the raven's burial can also be found in Jewish folklore.[6]

Stories of other nonhuman animals exhibiting signs of grief at the death of a companion abound, and they include chicks, cows, pigs and many other species.[7] Cetaceans and primates have been observed carrying their dead infants for various periods of time. For example, in 2010 Robin Baird from Cascadia Research reported watching a killer whale carrying a dead calf for over six hours. The calf still had the umbilical cord attached and it is not clear whether the calf was a stillborn or died soon after birth. When the researchers left in the evening the mother was still with the calf. On 24 September 2012 the Italian newspaper *Corriere della Sera* published on their website video evidence of a pilot whale engaging in an activity with a dead calf similar to that described above. Ecologist Paolo Galli observed that the physical degradation of the calf could indicate that the adult's interaction with the infant had lasted for several days and nights. More recently, Tahlequah, a twenty-year-old orca, who lost other children in the past few years, was observed carrying her dead daughter for an entire seventeen days.[8] Similarly, Biro and colleagues report on mothers in primate species carrying their deceased child. In one case a chimpanzee mother carried her infant's corpse for sixty-eight days: 'Over the days following death, the bodies swelled, then gradually dried out. All hair was lost, but body parts remained largely intact, encased in dry leathery skin.'[9]

---

5   Qur'an 5:31: 'Then Allah sent a raven, who scratched the ground, to show him how to hide the shame of his brother …'
6   Reported in Gutmann 1998: 'Looking up, he [Cain] saw the bird that had killed its fellow putting its mouth to the ground; and it dug [a hole] and buried the other dead one, and covered it with earth. Cain did the same to Abel, so that [his father] might not find him.'
7   e.g. Hatkoff 2009; King 2013.
8   Mapes 2018.
9   Biro et al. 2010, R351.

Bottlenose dolphins appear to react differently to sudden, unexpected death than to a death that has speculatively been labelled as predictable and that may represent a warning signal to the rest of the community. In one case, for example, a mother stayed with her dead newborn calf for two days, calling to, touching and lifting the infant's corpse above the water surface. In a second instance a two- or three-month-old dolphin, bearing bleach marks as a possible result of exposure to pesticides or heavy metals, was having difficulties swimming. Other dolphins tried to help the dying individual stay afloat but with little success. Death occurred an hour later. Contrary to the researcher's expectation based on past experience, the mother this time did not stay nearby, and instead, as soon as the sick dolphin died, the corpse was allowed to sink and the group left the area.[10] Of course, as Marino points out, it is difficult at this stage to explain these differences as ones reflecting expected and unexpected death since the reaction may reflect individual rather than categorical, species differences, and more research/observation data is needed to clarify patterns.[11]

Cases of an animal's grief ending with the grieving individual's own death are also known. At the end of the previous chapter we encountered 'suicidal' dogs.[12] Another such example is Damini, an elephant in an Indian zoo, reported by *BBC News* on 6 May 1999 as having starved herself to death after her friend had died in childbirth. A similar episode almost occurred at the Edgar's Mission Farm Sanctuary in Australia in 2010 when Daisy, a pig, died, and her close companion, Alice, lay on her grave for two days and nights, refusing to eat or move.[13]

Such accounts unavoidably stimulate the curiosity of scientists and non-scientists alike. Testimony to this is the growing interest in comparative thanatology[14] and the emergence of the new field of evolutionary thanatology, which may eventually aid bridging extant conceptual gaps concerning cross-species death and grief. For a long

---

10   Hooper 2011; see also Reggente et al. 2018 and Bearzi et al. 2018 for recent reviews of marine mammals' death-related behaviour.
11   Lori Marino, email, September 2019.
12   Preti 2007.
13   Pam Ahern, personal conversation, 2011.
14   Anderson 2016.

time, in fact – and to a great extent this remains the case today – the issue of nonhuman animal grief was not explored. Instead, it was avoided – in effect dismissed – by asking potentially unanswerable philosophical questions, such as: Do nonhuman animals understand death? Do they know they are mortal? Can they *really* grieve? These questions are often accompanied by explicit or implicit reminders of the 'dangers' of anthropomorphising. Such rhetoric gives ample space to translate the privacy argument – according to which conscious experiences can only be known by the individual who experiences them[15] – into negation of nonhuman animal grief. Additionally, the implicit message it conveys is that even if nonhuman animals might feel 'something' at the loss of a proximal subject, it obviously cannot compare to humans' grief, given that humans can answer positively to the questions above, or so it is assumed, while other animals cannot. This becomes particularly problematic when even apparent 'animal lovers', in the pursuit of 'intellectual rigour', explicitly though without any sound foundation, ignore the privacy argument by stating, for instance: 'They [elephants] can't anticipate death in the way we can, or imagine it as an abstract concept.'[16] This is incorrect; in reality, humans simply do not know. Other animals' grief, wildlife film-maker James Honeyborne feels the need to add, 'is *different*: it's *simply* about loss' (italics added). This statement, recurrent in discussions of nonhuman animal grief, fails to appreciate the potential affective power and magnitude of the feeling of loss. It also erroneously assumes that all humans will contemplate their own mortality when they experience loss, and even that all human survivors necessarily fear death or want to – and know how to – live following a loss. On the contrary, loss may, in fact, put human survivors at higher risk of mortality, including suicide.[17]

Human and nonhuman animal structural and functional brain comparability allows us to infer what we know about the raw human experience of grief to other animals. However, acceptance of a human-comparable grief dimension in nonhuman animals is often

---

15 Griffin 2006.
16 Honeyborne 2013, n.p.
17 e.g. Kaprio, Koskenvuo and Rita 1987; Erlangsen et al. 2004; Rostila, Saarela and Kawachi 2013.

hindered by introducing into the equation these philosophical side questions. An additional problem is that even though reports of nonhuman animal grief expressions and related rituals are not uncommon, overall they do remain sporadic, not only in the wild but also on farms, where the captives' feelings tend not to be a prioritised aspect of the operation.

The emphasis upon mourning practices above subjective experience in the consideration of death and grief – and in the case of nonhuman animals, for the recognition and acceptance of grief – appears to be the result of a history of 'caution' in the deliberation of questions related to subjective experience in all sciences, including anthropology. As anthropologists Richard Huntington and Peter Metcalf point out:

> The interpretation of emotional states presents special problems to anthropologists because the discipline focuses on culture and society, on communal ideas and corporate structures. The emotions within individuals are largely beyond our purview. We have learned that we must be cautious in attributing particular emotional configurations to members of other cultures.[18]

A certain measure of caution is indeed in place. This is not because, in the case of the feeling of grief, there would be any reason to doubt that members of other human (and nonhuman) cultures experience the intense pining and yearning at the core of grief. Rather, because the sensitivity to cultural diversity in the dealing with and expression of grief may determine our approach – when our intervention is requested or unavoidable – as helpful or otherwise. In the case of refugees who have suffered trauma and loss, for example, a deeper understanding of their cultural normativity informs approaches that would promote healing, as social anthropologist Harri Englund points out in a discussion on death, trauma and ritual in relation to Mozambican refugees in Malawi.[19]

---

18   Huntington and Metcalf 1979, 23.
19   Englund 1998.

The attentiveness to outer expressions of grief and mourning rituals in anthropology and social sciences has produced an extensive corpus that testifies to the diversity of practices surrounding death, meanings of death and coping with loss across human cultures. When nonhuman animal grief is discussed, however, this diversity is usually ignored and a non-existent human universality in relation to these questions is promoted. This reflects largely the position/understanding of the WEIRD population, an acronym introduced earlier in this book and coined by Henrich and colleagues for the article 'The weirdest people in the world?'[20] In this article they expose the problem of the widespread practice in (human) behavioural sciences of using samples from this small and, in their opinion, particularly unusual slice of population compared to the rest of the human species to make broad generalisations across the entire species. They review various domains, such as visual perception, spatial reasoning, cooperation, self-concepts, and other aspects that vary across human cultures. A similar phenomenon manifests in discussions of nonhuman animal grief. The question of the understanding of death and one's own mortality is a case in point.

In most human societies, including the West,[21] death is viewed as some kind of transformation, not as annihilation; therefore humans' understanding of death and their own mortality may differ in both kind and degree. The WEIRD's tendency to impose categorical divides obfuscates the discussion of nonhuman animal grief, particularly in the absence of specifications of what is meant when, for instance, the question of 'understanding death' is raised. Thus in considerations of nonhuman animals' understanding of death, a working definition of death – possibly one that is applicable beyond the discussant's own culture or society – would contribute to greater clarity, and potentially fruitfulness, of such discussions compared to the current unfounded generalisations. Such generalisations only serve to promote further separation of humans from the rest of the animal world. There is ample evidence to indicate that, like human animals, nonhuman animals 'understand', as in *recognise*, death as organismic annihilation with

---

20 Henrich, Heine and Norenzayan 2010.
21 See, for example, Destro 2009, 536.

consequent physical non-returnability (in the current shape and form). There is certainly unequivocal evidence of nonhuman animals' capacity to experience attachment and separation distress, as shown previously. When organismic annihilation affects an individual to whom an animal – human or other – is attached, grief follows.

It is, however, critical to distinguish the experience of grief, which is largely neurobiologically determined and clearly cross-species, from the expression of grief, which is heavily culturally informed. The absence of evidence of mourning does not necessarily imply absence of grief – an assumption that keeps plaguing discussions of nonhuman animal grief. The absence of mourning evidence may be a result of a number of factors. For instance, species-specific rules may dictate avoidance behaviour in relation to corpses, individual differences may also influence expression of grief, and if an observed animal does not appear to be grieving, it does not necessarily mean that he or she is not. Further, we are not close enough – both physically and in terms of knowing them intimately – to most animals to appreciate nuances in behaviour, facial expressions, body language and other factors that may provide cues for determining presence or absence of grief following a loss. Companion animals are an exception because we live with them and know them better, which in many cases enables us to detect the slightest of changes. Humans who have witnessed nonhuman animal loss in the intimacy of the domestic environment, or sanctuary, and humans who even deal with nonhuman animal grief on a regular basis, are well aware how deep and strong this feeling may run.

In May 2016 a veterinary ethologist contacted me in her search for opinions concerning her desire to commence research in canine grief using functional magnetic resonance imaging (fMRI). This technology measures brain activity based on changes of blood flow to various areas. She had been working with grieving dogs and felt the need for a deeper understanding of the phenomenon. While to various extents grief remains enigmatic, new research methodologies are helping clarify the mechanisms involved.[22] This may enable higher efficacy in aiding the bereaved, particularly when pathologies such as complicated

---

22   The results of the first functional neuroimaging research into the neuroanatomy of grief, authored by Gündel et al., were published in 2003.

grief develop. A greater insight into other species' grief, for example through observation of brain activity, would certainly be interesting and potentially helpful for both therapy and advocacy for other species. However, various ethical and methodological issues would need to be resolved before such an experiment could be carried out. In my response I pointed out three areas of concern: i) technical problems in using fMRI for non-invasive observation of negative emotion in nonhuman animals; ii) the question of the animal's understanding of death as physical non-returnability and the possible complications in establishing this for the purpose of the experiment; and iii) the potential breach of species-specific cultural norms in attending to grief and its expression. The rest of this chapter discusses these issues from a cross-species perspective.

fMRI and dogs

Traditionally brain scans of nonhuman animals have been conducted on either anaesthetised or restrained and immobilised individuals. Neither of these methods is suitable for testing cognitive function and emotional responses in nonhuman animals: anaesthesia for the obvious reason that the animal needs to be awake for such testing, while immobilisation would trigger negative emotional responses in the tested individuals and likely compromise the results, as pointed out by Berns, Brooks and Spivak, who have conducted fMRI studies on awake unrestrained dogs.[23] One of the goals of their 2013 study was to determine whether their previous results could be replicated, and the article also explains in some detail the training process of the canine participants. Two further studies based on canine fMRI were published in 2016: the first one, by Berns's group at Emory University, tested whether dogs prefer praise or food,[24] and the second one, by a Hungarian group from Eötvös Loránd and Semmelweis Universities, looked into dogs' capacities and modalities of segregating and integrating human words and intonational cues.[25]

---

23   Berns, Brooks and Spivak 2013.
24   Cook et al. 2016.
25   Andics et al. 2016.

The fMRI scanner is a noisy and generally highly unpleasant environment for humans and dogs alike. The dogs have to habituate to the scanner and lie completely still for a certain period of time. It takes between five and twenty sessions of basic training outside the scanner, explains ethologist Márta Gácsi from the Hungarian team in an interview with *The Washington Post*,[26] and an additional ten sessions inside the scanner. The duration of the training could be further impacted by the availability of the dogs' human companions to bring the dogs in for training as well as by the researchers' accessibility to an fMRI machine. The latter, however, ceases to be a factor if the research lab has access to a replica MRI scanner as in the case of Emory. Generally, with weekly or fortnightly sessions, the training is completed in two to three months. Training methods revolve around positive sentience with emphasis on social rewards (for example, praise) rather than food rewards, specifies Gácsi, even though the latter are also employed. Using the so-called model-rival training method with groups of dogs, the purpose is to bring the dogs to *want* to be part of 'the game': to go into the scanner and perform well. Dogs function partly as models showing others where to be and what to do but they are also rivals because the performing dog receives all the attention while others are ignored. In regard to dogs who have successfully completed their training, Gácsi explains in the interview:

> you can see in their eyes when a drop of water falls on their noses and they know, 'I cannot lick it.' It's really … I don't know what to say. They are not forced. They are asked. You can't imagine how happy they are at the end. They bounce to the others like, 'Okay, I did it! I did it!' We are really seeing that they are proud.

Aside from the ethical problem of the inherent instrumentalisation that any kind of training entails,[27] this method is clearly efficient for the research purposes designed and carried out, but it would be inadequate in grief research. To begin with – and setting aside the question of

---

26   Brulliard 2016.
27   In a species-normative environment, like humans other animals are also *raised* and *educated*, not *trained*.

the (in)sensitivity of putting a grieving dog through such intense and stressful training – the hyperactive social environment and emotionally positive nature of such training would certainly lead to compromised results if used in studying grief. The training period would also have to be substantially shorter than two months since, in the absence of complicated grief, after two months acute grief should recede and acceptance of and adjustment to the death begin to take its place. Further, of the attributes predicating successful completion of training of the canine individuals Gácsi emphasises trust in the human guardian and eagerness to please. Trust is a complex phenomenon that is built over time between individuals, and is constituted of a feeling of confidence that the other's actions will be characterised by positive effects or at least by the absence of negative effects on the individual, including the respect of personal boundaries. The usurpation of such trust in a grieving dog for a procedure that will not bring direct relief to the dog, and that may even increase emotional suffering, bears consideration. Furthermore it is questionable whether a grieving dog will be equally eager to please the human guardian, and importantly, as per the above discussion of trust, whether it is ethically sustainable to demand/induce such behaviour in a suffering dog.

If the above issues could eventually be resolved and dogs could be trained for grief research in a non-invasive way (if that is possible at all), other methodological problems would have to be addressed before such research could be carried out. For testing human subjects, images (for example, the deceased vs a stranger) and words (for example, grief-related vs neutral words) are usually employed as cues.[28] However, dogs are understood to be less visually oriented compared to humans; visual material may not be the optimal choice for cues. Instead, aural (the deceased voice) and olfactory cues may be favoured. The noisy environment inside the scanner makes wearing earmuffs a necessity. Rather than being an impediment to the employment of aural cues as I initially assumed, speakers can be incorporated in the earmuffs. These sound-attenuating headphones, as the researchers call them, were also used in the Hungarian study.[29] More interesting and certainly more

---

28   e.g. O'Connor et al. 2008.
29   Attila Andics, email, 20 December 2016.

complex would be the choice of olfactory cues. In this case the researchers would have to decide whether to use material (for example, hair, bedding, etc.) that bears the smell of the deceased individual from before or after their death. Ants have been shown to recognise a dead conspecific by the smell of triglycerides that the cadaver emits.[30] To my knowledge, canine modes of death recognition are unknown to humans. Anthropocentric scepticism aside, nonhuman animals can certainly distinguish a dead organism from a live one – an essential survival tool – and in the case of canines a strong olfactory component is likely involved. The choice between pre- or post-mortem cues (for example, hair) may affect the experiment's results, perhaps particularly in relation to the brain's reward centres. In this case extreme caution would be required in the interpretation of results since activation of reward systems may lead humans to automatically assume that the animal does not understand the companion's permanent departure whereas the animal may in fact be experiencing complicated grief. As noted in the previous chapter, the brain reward system has been shown to activate when testing humans with complicated grief.[31] First and foremost, however, researchers would have to ensure the dog has been given the opportunity to establish the death of the conspecific instead of assuming the possibility of return. Thus the human guardians would have had to expose the dog to the corpse and let him/her interact freely with it if the dog chose to do so.

This brings us to another question: the question of canine culture surrounding death and grief, and what this may signify for the ethics of the experiment and possibly for the results. As we will see later on, observations of human behaviour in relation to death and grief across cultures show a diversity of culturally informed approaches ranging from close interaction with the corpse to virtually no interaction, and from expressive manifestations of grief to stoic attitudes where displays of emotions are suppressed. It is possible, in fact probable, that nonhuman animals have developed various species-specific, and

---

30   Reported by Wilson 2009, and Hirshon n.d. Some ants, such as the Argentine ant, rely instead on the disappearance of cuticular chemicals emitted by live ants (reported in Gonçalves and Biro 2018).
31   O'Connor et al. 2008.

possibly also society-specific, rules or principles in the approach to cadavers. Humans, in our lack of knowledge and understanding, may breach these rules by forcing re-exposure for research purposes. For canine companion animals, anecdotal evidence testifies to a variety of responses that may reflect individual differences but may also depend on the quality of the relationship between the dead and the surviving canine. Dogs may show little or no interest in the corpse itself, as in the recent case of Minka when she lost Daz, her canine companion of thirteen years. Minka paid no attention to the corpse itself; however, following Daz's burial, Minka lay on the grave and refused to leave the site, explains her human guardian, psychologist and educator Clare Mann.[32] Similarly, when Bobbie, my parents' dog, died, I invited his two canine friends from the neighbourhood to witness his corpse, thinking they should know why they would not be seeing and hearing him in the future. They were not close friends, but in the microcosmos of the domestic space, which is shared with companion animals yet inherently dominated by humans, I believe the least we can do is to encourage participation at all levels. This includes avoiding disappearances of subjects and the breaking up of relationships (close or less close) without attempts to explain the situation to the potentially affected others (in this case the neighbouring dogs) as best as we can. The first surprise came from the human guardians of the neighbouring dogs, who were reluctant to expose their canines to their deceased neighbour without providing an explanation for their hesitation. After some persuasion they let me lead their dogs – separately, not both dogs together – to the corpse, which was lying on the ground ready to be buried. The second surprise came from the dogs themselves: both dogs stood at a distance of about a metre from the cadaver, sniffed it from that distance, seemingly untouched, refusing, however, to approach the corpse despite my encouragement to do so, and indicating clearly that they wanted to leave the scene – the dogs were not restrained but they tried to oblige me. Another surprising fact is that three years on, at the time of writing, the deceased dog's favourite toy, which was placed on top of his grave in the garden at the time of burial, remains intact even though a new puppy was adopted six months later who, along with

---

32   Clare Mann, personal conversation, 2017.

other canine neighbours, has always had full access to the grave area and the toy. More recently Charlie died, our canine companion for close to fourteen years and the sheep's companion for seven. In the last stages of his battle with cancer he was quite weak, and the sheep, on our walks together, were used to me picking him up when he became tired and carrying him in my lap. When he died, I wanted the sheep to know. I approached them with Charlie in my arms. They came to greet him but soon realised something was not right. They sniffed him, and jumped off. One by one. It was clear that they recognised the olfactory cues signalling death, whatever death may signify in their interpretive world. We buried Charlie under the lemon tree in the vegetable garden. Every year in autumn we let the sheep into the garden area to help us clean it up. One of their favourite treats is the lemon tree, which we always have to protect so they do not eat it bare. This year, however, and reminiscent of the above toy account, the sheep did not touch it.

All things considered, it remains unclear how exposure to visual, auditory, olfactory or other cues in a dog (or other nonhuman animal) that appears to be grieving for research purposes may affect the studied animal – would it contribute to an intensification of emotional pain? Would it scare them? Would it confuse them? – and as a consequence how these unresolved questions may impact upon the research results. Let us now turn to the complex question of understanding/recognising death.

Understanding death

The chapter on attachment and loss gave an insight into the processes involved in the development and maintenance of attachment relations, which also helps us appreciate how separation distress can be equally painful and debilitating (to the point of death in some cases) for human and nonhuman animals alike. Yet to feel grief as death-related loss, one would have to have some kind of awareness of death. As already indicated, the commonly asked question 'Do nonhuman animals understand death?' is in effect a composite of two sub-questions or sub-issues. The first sub-issue concerns their ability to understand what, for want of a better term, I call the physical non-returnability of the (dead) subject (and physical non-returnability is what I refer to when I use the term 'death' in this discussion). The second concerns their ability to understand their own mortality: knowing that they can die and that one day they will.

It is not uncommon for humans to question nonhuman animals' capacity to recognise death when encountered. Such questions are often raised, for example, when nonhuman mothers are observed carrying their dead infants.[33] However, it seems more credible, given the regular occurrence of death and animals' regular experience of dead organisms, to consider the awareness or recognition of death as part of their phylogenetic knowledge (the organism's inherited long-term knowledge), which subsequently becomes complemented and fine-tuned by life experience and encounters with death.

Phylogenetically based recognition of death and its benefits for survival would, in fact, make good evolutionary sense for both so-called prey and predator species. Further, even though some human societies are notoriously expressive around death and corpses, in the rest of the animal realm death is also better thought of as a phenomenon that does not belong solely to the individual affected (and close others). In many respects, and particularly in group-living animals, death as such could be considered a social event in the sense that something that happens to a member in the community potentially affects other members. Therefore the event or phenomenon warrants attention. For instance, a death may impact upon the current social organisation of the community. It may also supply important information for the community, such as (depending on the mode of death) signalling the presence of a predator in the neighbourhood or a contaminated environment. Social animals will simply pay more attention to social things, and as we know from cognitive studies, they will perform better on tasks that engage their social faculties.[34] It is not too farfetched to consider that in all likelihood various animal species will have developed distinct ways of establishing that an organism is dead as opposed to alive.

Ants are interesting in this regard. Ants obviously recognise death, as they remove their dead conspecifics from the nest (a phenomenon

33   e.g. Biro et al. 2010; Sugiyama et al. 2009.
34   Marino 2017.

known as necrophoresis) and pile the corpses in what we could call ant cemeteries. As mentioned before, ants recognise death when they detect the smell of triglycerides. Wilson reports watching a live ant whom he had covered with the death-chemical being carried to the cemetery by other ants who believed her dead, ignoring her other signs of life, such as limb movements.[35] Considering ants' olfactory dependence, this comes as no surprise. Creatures with an underdeveloped olfactory consciousness, such as humans, rely on other information to establish an organism's death: immobility, for example, may be one of the first signals. In fact, for most of human history, sociologist Allan Kellehear notes, death was determined based on behavioural observations rather than assessment of physiological states, and the process was gradual: cessation of movement was followed by observations of other signs, including stiffening of the body, colour changes and changes to the eyes, and finally putrefaction or maggot invasion.[36]

Today the general criteria for death in a hospital situation are 'irreversible cessation of circulation and respiration or irreversible brain function (whole brain, that is, cerebral hemispheres and brain stem; or brain stem alone)'.[37] However, Kellehear also reports that these criteria are not considered reliable for neonates. This may explain the occasional error such as in the case of the Australian baby who was proclaimed dead when he was not.[38] In 2010, Emily and Jamie Ogg were born premature at a Sydney hospital. Emily was fine but Jamie was not breathing. Following unsuccessful revival attempts by medical staff, it was the nursing method known as 'kangaroo mother care' or 'skin-to-skin care' that enabled the baby to live. The parents wanted to say goodbye to their dead baby and the nurses placed Jamie's body across the mother's bare chest. A few minutes later, Jamie started to show signs of life, which were becoming increasingly more pronounced. Jamie was alive. Kangaroo mother care is a useful technique in places where incubators are not readily available, and obviously elsewhere. It might constitute part of the phylogenetic knowledge of the previously mentioned carrying

35 Wilson 2009.
36 Kellehear 2008.
37 Kellehear 2008, 1535.
38 Inbar 2010.

nonhuman mothers, considering the vital importance of such interactions within the mother–infant dyad. Based on the available science, Dr Lisa Eiland from the New York City Weill Cornell Medical Center, commented on the Australian case:

What's important is the warmth that the mother provides and the stimulation that the baby may have received from hearing the mother's heartbeat. So those are all things that may have helped the baby in terms of going down the path to living as opposed to the path of death.[39]

The occasional prolonged carrying of the dead infant to the point when the baby mummified, observed in primates, and other signs of reluctance to 'let go' that other animal species may display, can be viewed as a sign of the bereaved individual's psychological distress and inability to accept the death rather than a sign that nonhuman animals might not recognise death.

H. Clark Barrett and Tanya Behne, studying human children's recognition of death in a cross-cultural context, propose the existence of an early-developing core architecture, which they call 'the agency system' and which helps children discriminate between living organisms and dead ones.[40] The researchers found that based on the agency criteria (agent being defined as an 'object capable of acting in a goal-directed fashion'[41]) children understand the concept of the irreversibility of death by age four. Previously it had been assumed that children could not grasp irreversibility until they are six or seven years old. Given the latter figures and the widespread notion – redolent of conceptual and methodological problems – that no other animal reaches the cognitive level of a five-year-old human child, John Archer in considering grief from an evolutionary perspective dismissed the possibility of nonhuman animals' awareness of the irreversibility of death altogether.[42] However, Barrett and Behne correctly point out

---

39   Reported in Inbar 2010, n.p.
40   Barrett and Behne 2005.
41   Barrett and Behne 2005, 95.
42   Archer 2001, 265.

that children might develop an understanding of death as cessation of agency before the age of four, and that studies of this kind, including studies of the theory of mind, are always problematic as they are dependent on other acquired skills, such as language.

In Chapter 4 we will consider the implicit aspect of agency as a quality of phenomena that materialises through an individual's response to it. Animals can engage with such implicit agencies on a purely experiential level without the need for cognitive elaboration. However, a substantial amount of an animal's life is focused on explicit agencies, both in relation to individuals from one's own community and other communities, including of course potential predators. Agency detection is, in fact, one of the most vital capacities in the animal realm, as is the capacity to predict, to various extents, the agents' intentions. As Benson Saler points out discussing Stewart Guthrie's take on anthropomorphism, we are 'not only disposed to be sensitive to the possibilities of human-like intentional agents in the environment, *but we are given to over-detecting them*, and doing so may ultimately be in our best interest'.[43] The same applies to other animals, whose lives and relations are dependent on the ability to detect and interpret agency and intentionality.

In 2013[44] I introduced 'intra-zoomorphism' as an umbrella term for the various kinds of 'ego-morphising'[45] animals engage in, which includes humans with our ego/anthropomorphising. Intra-zoomorphism encompasses both inter- and intra-species 'morphising', including, for example, inter-cultural 'morphising', inter-gender 'morphising', and inter-personal 'morphising'. The inevitability of relationships between different 'others' compels the subjects involved to formulate ways of negotiating these relationships. Of course, humans are not the only animals to encounter otherness. If we maintain the position that we can only see the world through our own 'eyes' – without denying ourselves and others the necessary space that enables more or less successful interaction – the same must be true for other species. This capacity of identification and projection may take the form of nurturing bonds

---

43 Saler 2009, 46; Guthrie 1980.
44 Brooks Pribac 2013.
45 Projection of one's characteristics as an individual or group to other individuals/ groups.

between different animal species. In *Unlikely Friendships* Jennifer Holland documents a number of supposedly unusual inter-species relationships, such as the friendship between an elephant and a sheep or that between a bear and a cat.

A less unusual, in fact very common, example of intra-zoomorphism can be observed in the communication between nonhuman animals and their companion humans in domestic situations. When Charlie began nudging me around 4 pm every day it was legitimate for me to assume that he was doing so in the conviction that I would understand that it was time for his walk. Stuart Watt refers to his cat's gestural language while interacting with him as a member of the cat's social group as a possible instance of *ailuromorphism*,[46] while Bekoff coins the term 'dogomorphism'[47] to refer to canine self-projections such as those demonstrated in a study of dogs and dog-like robots.[48] In this study dogs only considered the robots covered in fur as potential social partners. Before rushing to unjustified conclusions, it is worth pointing out that dogs are not the only animals who might be misled by robots. A 2007 study revealed that the human mirror reaction is activated without significant difference by the sight of both human and robotic actions.[49] Mirror neurons, which enable this reaction,[50] are brain cells that make one salivate while watching somebody eating an ice-cream, and that deliver a heartbreak at the sight of a mother cow trying to prevent her baby from being taken away so that the ice-cream could be produced in the first place. Mirror neurons fire during both the execution and the observation of a performed action. Nicknamed 'empathy neurons',[51] mirror neurons might provide the biological base for the tendency to ego-morphise.

Of course not all relationships are convivial, and for these relations, such as the relation between a predator and a prey individual/group,

---

46 From *ailouros* – Greek for 'cat'. Watt 1998.
47 Reported in Horowitz and Bekoff 2007.
48 Kubinyi et al. 2004.
49 Gazzola et al. 2007.
50 These neurons were first discovered in the early 1990s in macaque monkeys but the electrophysiological evidence for their existence in humans is a more recent occurrence (Keysers and Gazzola 2010).
51 Ramachandran 2006.

agency and intention detection are equally important. An attack could be as deadly for the prey as it could be for the predator, hence the predator has to carefully evaluate the situation and be fully aware of and in control of their own mind and body state, as well as try to predict that of the prey. In fact, '[i]f a hunter is not aware of the state and condition of her body and mind, then she compromises her ability to judge accurately whether a particular action constitutes a threat or benefit, death or food'.[52] The same applies to the targeted animals who will have to decide whether to attempt escape or defence. The latter may involve, if the targeted animal is part of a group, additional evaluations of agency and intentionality, namely the likely response of their conspecifics and the potential success or lack thereof. 'Group defence in horned animals such as buffalo, eland, bison or musk oxen can,' Broom points out, 'prevent completely the death of group members when there are attacks by hunting dogs, hyenas or wolves.'[53] However, an uncoordinated response and/or misinterpretation of the circumstances by an individual or group may lead to an even greater tragedy than attempting escape.

In any case, detection of goal-oriented agency is so paramount in the animal world that by analogy animals must also detect, recognise its absence, and more specifically the irreversibility of such absence in the case of death. Being able to discriminate between sleeping offspring and dead offspring or between a sleeping lion and a dead lion, if I am his potential prey, informs behavioural choices and energy and resource investment. The benefits associated with success and the costs associated with failure, Barrett and Behne note, have shaped the evolution of agency-detection systems, present in all animals.[54] Different species may have different ways of recognising death, and sometimes, it seems, even predicting death, as in the case of Oscar, a resident cat at the Steere House Nursing and Rehabilitation Center in Rhode Island: when Oscar curls up beside a patient, the medical staff know that death is imminent, allowing them to notify the family who may wish to be present as the patients take their

---

52   Bradshaw 2017, 128.
53   Broom 2003, 53.
54   Barrett and Behne 2005.

last breath.[55] In any case, the capacity for death recognition, if not necessarily prediction, is most certainly shared across species.

The question of nonhuman animals' awareness of their own mortality, on the other hand, is of a far more speculative nature: it would be as incorrect to posit that nonhuman animals are unaware of their mortality as it is to claim they are aware of it. While the question may never be resolved, the increasing evidence for nonhuman animals' capacities for bi-directional mental time travel (remembering past events, and anticipating and planning the future),[56] accompanied by their capacity to experience learned (as opposed to only phylogenetic) fear, and the fact that historically death has been a recurrent phenomenon in most animals' lives and environments, suggest that the possibility of such abstract thought in nonhuman animals should not be automatically ruled out. Awareness of death would fall under the category of inevitability/universality of death in measurements of the so-called 'mature' concept of death,[57] a notion that is both anthropocentric and potentially ableist. The inevitability/ universality criterion along with the causation criterion whereby the individual understands that ultimately the cause of death is 'a breakdown of bodily functions'[58] are commonly used to reject the idea that nonhuman animals (along with human children) may have a fully developed understanding of death. While these two aspects of understanding death may represent interesting investigative topics, they are not directly relevant for the discussion of grief.

Humans' awareness of mortality, for instance, appears to play a lesser role in grief experience than is often implied when considering nonhuman animal grief. Indeed, lack of awareness of mortality certainly does not preclude the feeling of grief. The human notion of mortality, as previously noted, hardly bears the unanimous consensus that it is often credited with in discussions of human versus nonhuman animals' grief. Arguably, the numerous humans whose religious feelings give them

---

55   Dosa 2007. Oscar's sense of death probably results from his capacity to detect subtle chemical changes in a dying body (Pierce 2013).
56   Roberts 2007; Clayton and Dickinson 2010; Cheke and Clayton 2010.
57   Longbottom and Slaughter 2018.
58   Longbottom and Slaughter 2018, 2.

internal certainty of a new, even better, life after physical annihilation, do not appear to understand death and their own mortality in the way that secular Western thought postulates when raising arguments against nonhuman animals' grief. Is the grief experienced by these supposedly immortal humans at the loss of a proximal subject less acute than the grief of those humans who believe that their entire persona dies when the body capitulates? Does the certainty some humans hold of a reunion with the beloved after the physical death preclude grief? Should these potentially immortal humans be denied the full extension of the feeling of grief based on their apparent certainty of eternal life? If the answer is negative – without necessarily dismissing possible variations on the theme – and in consideration of the potential impact of separation distress per se, denying nonhuman animals a human-comparable grief dimension appears premature to say the least.

Therefore nonhuman animal grief may be more appropriately considered within a broader cross-species and cross-cultural context, in which nonhuman animals would represent another (obviously very large and diverse) cultural group that is potentially not aware of their own mortality. It is on this basis that nonhuman animal grief can begin to be assessed without sinking into unnecessary and potentially damaging speciesism.

Another issue to consider in grief studies is what I refer to as the ease-of-identification fallacy, which is twofold: first, dismissing from consideration animals with whom humans cannot easily identify; and second, dismissing from consideration animals who do not exhibit grief-like behaviour that is identifiable as such to the human observer. Invertebrates represent an interesting case in question. Without necessarily attributing the experience of grief to invertebrates, it would be equally erroneous to *a priori* dismiss this possibility on the basis, for example, that expressions of grief have not yet been identified in invertebrates, or that science has not yet officially confirmed their capacity for attachment. Some invertebrates are very gregarious animals and highly attuned to the life of the collective. Wasps, for example, pay close attention to, and remember, social interactions among different individuals even when they are not themselves part of the game;[59] bees

---

59 Tibbetts, Wong and Bonello 2020.

are capable of emotional responses (like stressed humans, stressed bees tend to be pessimistic and respond negatively to an ambiguous stimulus);[60] while the story of ants liberating a nest-mate (but not unrelated ants) from a nylon snare she was trapped in, running the risk of harming themselves, demonstrates not only cognitive and behavioural complexity,[61] but possibly also attachment or altruism.

Different species' traditions could, in fact – for various reasons, including sanitation – explicitly prohibit any kind of interaction with cadavers. Certain human societies, for instance, forbid uttering the name of the recently deceased person.[62] There is no guarantee that equivalent and perhaps more stringent practices – that is, cultural lack of acceptance of any kind of expression of grief that humans might recognise – do not exist in other animal groups. Such absences do not necessarily indicate a lack of grief, just as the lack of vocalisations in injured animals is no guarantee that they are not in pain – an assumption that might lead humans to believe that mulesing, for example, is not painful because lambs tend not to cry during the procedure.[63]

Since when attempting to assess the presence or absence of grief in nonhuman animals we have to rely largely on the behavioural aspect, grief can easily be missed, especially in animals we do not share our homes with and hence do not know intimately enough. What we do know, however, is that animals in the wild and animals on farms and in other captive settings (when given a certain level of freedom) do form meaningful bonds, and when the bond is deep so is the scar when death, or other forms of separation, break the bond. What we also know is that in the hypothetical case that a shared human language was not available and we had to rely exclusively on behavioural cues in detecting and interpreting grief expression in some other human cultures, the picturesque fabric of such expressions or lack thereof may leave us puzzled, lead us to completely misinterpret the behaviour and motivation – a risk archaeologists run when trying to interpret material

---

60   Mendl, Paul and Chittka 2011.
61   Nowbahari et al. 2009.
62   Frazer 1922; see also Steward 2013 reporting on a breach of the naming taboo in relation to a dead Australian Aboriginal activist.
63   Broom and Fraser 2007, 63.

culture without information on ideological and social aspects[64] – and possibly question their capacity for grief.

Human diversity in grief expression and repression

We will now take a quick look at some of the human customs and culturally informed management of physical and social pain that differ substantially from Western approaches. These insights call for additional caution in considerations of nonhuman animal grief, particularly when discussants feel the need to compare other animals' expressions of grief or lack thereof with 'ours' (human).

In 1997, Parkes, Laungani and Young brought out the book *Death and Bereavement Across Cultures*, addressing mourning practices across major world religions and a number of small-scale cultures to help the medical profession be of better service to immigrant communities. A lack of understanding of different attitudes to death in human societies can inhibit Western understandings of differing practices in these communities. An Egyptian woman complying with the culturally prescribed lengthy period of mourning for her dead child,[65] for instance, could quickly be diagnosed with a pathological condition by Western clinicians, while a Chippewa mother who might show no apparent grief at the death of her baby could easily be seen as a psychopath.

Apart from the obvious benefits for immigrants, the recognition of the diversity of attitudes towards life and death, and of expressions of grief within the human species itself, along with awareness that socio-natural environments may contribute to such diversity, are fundamental for the discourse of grief from a cross-species perspective. Widespread overgeneralisations concerning what humans supposedly do and what nonhumans do not, or vice versa, hinder conceptual resolutions and further promote the human–nonhuman divide. The inclusion of human cross-cultural and cross-societal[66] contexts into

---

64 Bloch 1981.
65 Rosenblatt 1997, 41.
66 Attitudes and behaviour are not always culturally informed; they could be influenced by economic, environmental and other factors.

the discourse, on the other hand, humbles our judgment, sharpens attentiveness and helps build a stronger theoretical framework from which to explore grief and mourning in other species.

Delayed personhood

We begin by considering so-called 'delayed personhood' in human societies, which may to various extents serve as psychological protection for the mothers and other family members in environments where infant mortality is high. Many nonhuman animals, free-living and captive, live in contexts of high infant mortality or other forms of infant loss and may have developed parallel defence mechanisms. This may result in absence of grief expression; however, as we learn from human societies, this process of postponed attachment is neither automatic nor does it guarantee the absence of grief. In this section, we will also look at examples of grief expression and mourning from selected human cultures around the world and see how dramatically such practices may differ within humanity itself. The purpose of this section is not to provide answers concerning nonhuman animal grief; rather, it is to illustrate that behavioural cues may be highly misleading in assessing nonhuman animal grief, particularly when we do not know the individual intimately. This has obvious implications for scholarly rigour, but more importantly it may have severe damaging repercussions for nonhuman animals since humans, including (or particularly) scientists,[67] continue to dismiss either grief itself or

---

67 One such moderately recent example concerns a male kangaroo interacting with a dead female kangaroo's body (Formosa 2016). Four photographs of this interaction, taken by Evan Switzer, attracted media attention. One of them, for instance, shows the male kangaroo holding the dead female's head while her joey stands beside her. There was general popular agreement that we were witnessing grief expression. An 'expert' was then summoned – a senior lecturer in veterinary pathology – to provide his unbiased scientific opinion. He based his opinion on these same four photographs, presumably having also read Switzer's account. Echoing the old school's view, discussed in the first chapter, according to which the modus vivendi in the nonhuman world is violence and reproduction, the expert concluded not only that attributing grief to this individual was 'naïve anthropomorphism', but also that, since competition between males can be fierce and males can be violent in their harassment of

human-comparable levels of grief in nonhuman animals, consolidating speciesism.

In Chippewa culture mothers are urged to abstain from weeping at an infant's death to prevent the grief being passed on to the next child.[68] This position finds scientific support in the reality of trans-generational stress and trauma transfer; however, it does not necessarily prevent the mother's grief.

The discouragement of mourning for the as yet unnamed child is present also in Tongan culture and instantiated by the older women reminding the bereaved mother that she should silence her sorrow as the child was merely a ghost.[69] This phenomenon, known as delayed personhood but which would be equally legitimate to term delayed attachment,[70] has been observed in human societies with high infant mortality and a generally adverse environment, and refers to the practice of considering infants not fully human due to the uncertainty of their survival and fulfilment of the expected role. The emotional distancing, which sometimes extends to delaying naming the infant for a certain period, aims at protecting the family from the burden of grief.[71]

An acute and rather disturbing version of this phenomenon developed in Alto do Cruzeiro (Crucifix Hill), a shanty town in Northern Brazil, populated by rural migrants forced off their plots by large landowners who were accumulating land for sugar cultivation.[72] When Nancy Scheper-Hughes first visited the Alto as a Peace Corps volunteer in 1965, she was puzzled by the frequency of the ringing of bells. She inquired with her host only to be told that it was nothing,

---

females for mating purposes, it is not just that this male could not possibly have been grieving, he may have actually killed her (for a discussion and links to media coverage see Brooks 2018 and Brooks 2020). A less biased opinion would have admitted the possibility that the kangaroo was grieving without preventing the scientist suggesting other possible scenarios.

68   Lancy 2014, citing Hilger 1951.
69   Lancy 2014, citing Reynolds 1991.
70   The term 'delayed attachment' appears to be used also to refer to situations in which mothers want to attach to the newborn but are unable to. This is not how the term is used in this chapter.
71   Lancy 2014, citing Bugos and McCarthy 1984.
72   Scheper-Hughes 1989.

'just another little angel gone to heaven'.[73] She then returned in the 1980s as a medical anthropologist to examine the phenomenon of these angelinos (little angels) more closely, and recorded her findings in some detail in a 1984 article. Following three more trips to the area, her Brazilian experience culminated in the book *Death Without Weeping: the Violence of Everyday Life in Brazil.*

Like Tongan mothers, Alto mothers at the time also had to protect themselves emotionally by practising delayed attachment. High numbers of pregnancies accompanied by high child mortality rates due to economic constraints with limited access to clean water, quality food, medication and time for nurture, forced these women to develop this survival trait to protect their own wellbeing and that of the surviving children. In her study, Scheper-Hughes recorded an average of 9.5 pregnancies per woman; 1.4 of these pregnancies ended in abortion (spontaneous or induced) or stillbirth, and 3.5 children per woman died between birth and the age of five. At the time of the study there were 4.5 living children per woman, some of whom were weak and would probably never reach the age of five.[74]

As discussed in Chapter 2, intimacy – those vocal, olfactory, tactile exchanges between the infant and the caregiver – is as important for survival, wellbeing and the emotional, cognitive and psychological development of the child as proper nutrition and medical care. The absence of such intimacy may have contributed substantially to the infants' high mortality rates. However, even if the Alto mothers had been aware of this (and perhaps they were, at least on a phylogenetic level), there was little they could do since they were unable to take the infants to work (to the plantations, to the houses of the rich where they worked as maids, etc.). On the other hand, if they had been able to carry the infant around as many primates (including humans) do, delayed attachment would have been compromised since proximity facilitates attachment. In this case mortality rates would likely have remained high due to food scarcity but with little to protect the mothers' psyches and their exacerbated grief.

---

73   Scheper-Hughes 1989, 8.
74   Scheper-Hughes 1984.

Further, the miserable conditions with frequent deaths and a generally hard life in the Alto community gave rise to a culture that favoured some children over others. Those who appeared more active, lively and showed a 'readiness' for life, as Scheper-Hughes notes, were offered more support compared to the inactive, passive, quiet ones.[75] The latter not only suffered from undernutrition and malnutrition, which ran rampant in the community,[76] their chances for survival were further jeopardised by gradual neglect. In fact, the community believed that it was better, for the child and the mother, that innately or constitutionally weak infants, infants born with deformities, frail and, generally, those who appeared inept for life, died sooner rather than later, since it was believed they would not be able to defend themselves in life anyway.

Letting individuals die, or purposefully killing them, to allow others to live (or live a better life) may be less of a taboo in the real life of the human species than it is in the WEIRD population's imagination. Practices including but not limited to delayed attachment (and all that it may entail) and senicide (the killing of older people) point to heterogeneity within the human species itself, which often gets overlooked in discussing nonhuman animals' lives, including grief. Awareness of this diversity in humans may help us appreciate that when other animals behave in similar ways they too may be forced to do so by circumstances rather than acting out of lack of love and concern for their conspecifics. We know that some animals (human and nonhuman) may leave the elderly and weak behind, but others do not, opting to look after them instead. Certain environments and 'lifestyles' may be more favourable for extended care of individuals in need than others.

Senicide – at the other end of the line from delayed personhood – became the focus of discussions and moral judgments in 2010 when the Indian weekly *Tehelka Magazine* brought thalaikoothal to public attention. Referred to as 'mercy killing' by the locals, thalaikoothal in essence describes the killing of elderly people by their children when

---

75 Scheper-Hughes 1984.
76 The mothers, undernourished themselves and needing energy for work, refused to breast-feed, believing it would deplete their bodies of the little strength that was left, and relied instead on the milk formula available, which was a poor substitute for the mother's milk.

the latter feel that their situations do not allow them to look after their elderly parents properly, and/or the caring for the parents would endanger their own and their children's survival.[77] The seeming ease and lack of emotional involvement with which senicide is carried out, or the purposive detachment in relation to infants in other societies, or the reluctance to seek medical care for people when such is needed but it is too expensive, the outcome uncertain and hence the unwillingness to spend money on someone who will probably die anyway,[78] gives little indication of the actual emotional impact of such acts upon those who are induced by circumstances to perform them. Cultural outsiders are likely to regard these actions with horror while being (or choosing to be) completely oblivious of the environmental – including social – pressure surrounding the decision.[79]

Returning to the Alto community, there is little, if any, public display of grief when these babies die. If the mother weeps she is chastised by others and reminded that it is better this way, they died because they were meant to die, and she should abstain from weeping because 'her tears will dampen the wings of her little angel so that she cannot fly up to her heavenly home'.[80] Acceptance of death and filicide through selective neglect in this Catholic community was facilitated by the deep belief in God and heaven. 'Who could bear it, Nanci, if we are mistaken in believing that God takes our infants to save us from pain?' Scheper-Hughes was told by an Alto woman.[81]

In sum, the reality of frequent loss gave rise to delayed attachment as a defence/prevention mechanism, while the religious framework along with social support provided coping mechanisms for these women who ultimately had to choose 'life over survival'[82] if they were to survive at all. Without such defences they would have easily been

---

77 Shahina 2010.
78 Van der Geest 2004, 902, discussing the Kwahu people of Ghana.
79 The WEIRD population got a taste of the situation in early 2020 when due to the COVID-19 pandemic the possibility of senicide became very real in our world too, see e.g. Brooks Pribac 2020.
80 Scheper-Hughes 1989, 16.
81 Scheper-Hughes reported in 2013, 28.
82 Bradshaw 2013.

swallowed by grief, which in its complicated form can lead to physical and/or mental breakdown.

Nevertheless, absence of grief expression does not necessarily equal absence of grief. Nor is delayed attachment a mechanical process that all mothers who may benefit from it adopt automatically and with ease. The practical benefits reasoned in support of delayed attachment may clash with the mother's emotional urge to attach, hence the capacity for delayed attachment may to a large extent depend on the individual's psychological constitution, personal history and social context.

The baby is an inevitable part of the mother's life well before birth. For humans, for example, that is between 253 and 303 days on average. During this period the baby is fully dependent on the mother, responding to her physical and psychological state as well as auditory input from the external environment, all of which affects the developing foetus, setting the foundations for further development after birth. Simultaneously, the mother inevitably responds to the foetus or foetuses on a purely physical level (not all of which is necessarily pleasant) and to various extents also on the psycho-cognitive level. For instance, the moment a human mother decides to keep the baby when she has the option to choose otherwise, it would be legitimate to assume that this conscious choice will also have subconscious ramifications with an underlying hope, perhaps even belief, that the baby will survive. This adds complexity to the phenomenon of delayed attachment and can to various extents inhibit the capacity for such in different individuals.

Another complicated aspect of delayed attachment is that at a certain point the mothers (the parents, family) have to let themselves begin to attach to the child, given that the Alto mothers (and most certainly also mothers from other communities where delayed attachment is encouraged) do end up loving their children like any other mother. Deciding when it is safe for the mother to do so may be challenging. Nailza de Arruda, Scheper-Hughes's host mentioned earlier, lost all her children; they died unnamed a month or two following birth, and she could barely remember them, except for one, Joana. Unlike the others, Joana was properly baptised when she was a year old; entrusted to the presumably powerful protection of Joan of Arc, she was expected to live, but she did not. She joined the little angels the following year, leaving the mother in grief – a grief that was considered 'inappropriate', 'a kind of madness' with the mother often sitting in front of Joana's photograph and talking to her passionately, shifting from imploring to recrimination, as she was trying to come to terms with the death but could not.[83]

In essence, cultural normativity in grief and mourning matters may to a greater or lesser extent protect individuals from pathological outcomes. Delayed personhood/attachment along with the discouragement of public expression of grief for the angelinos is certainly oriented towards this aim. However, and relevant for the discussion of nonhuman animal grief, these communities also show that behavioural cues in attempting to detect grief (the principal method used with nonhuman animals) are far from reliable.

Breeding machines?

To revert to the nonhuman animal world briefly, gestation periods in nonhuman animals vary between species and between subspecies within a species: for instance, the gestation period for sheep is between 144 and 152 days, for goats between 136 and 160 days, for pigs between 101 and 130 days, for cows between 279 and 287 days, for horses between 329 and 345 days, for whales between 365 and 547 days, for elephants between 510 and 730 days, and so on.[84] Like human females, other animal females experience hormonal changes and other symptoms (pleasant and unpleasant) used by the body to alert them to their new condition. Conversely, the developing foetus, fully dependent on the mother, absorbs nutrients, both dietetic and psychological, that the immediate environment has to offer. For the duration of the gestation period the mother and the infant(s) are in continuous implicit communication with each other, building towards what in a normative and plentiful environment would be called the 'happy event'. But there is not much normativity left for many nonhuman animals, particularly those in captivity.

---

83   Scheper-Hughes 1989, 8.
84   Infoplease 2000–17.

Maternity has become a function in the system of exploitation of nonhuman animals; it is, in fact, the most important function as it enables the system's very existence. The new role of these female bodies is massive production of babies, in order for the babies to be taken away from the mother and used for human interests: eat their bodies, drink their milk, (ab)use them as entertainment in circuses and zoos, experiment on them, and so on. The delicate process of a life emerging and developing has been fine-tuned over a long evolutionary history to enable optimum psychobiological growth in a given socio-natural environment, and increase chances of survival and quality of life. This process, through which the mother–infant bond evolves as a necessary constituent in the formative period spanning from conception to independence, is appropriated by an animal species that considers itself cognitively and ethically most advanced compared to all others species, and turned into a mere commodity.

I consider myself fortunate for living with one of the most unprivileged species and getting to know individuals from other equally unprivileged species through the occasional fostering or help at sanctuaries for rescued 'farm animals', individuals whom most people in our society only 'meet' through individual body parts served on their plates. While personally I have abstained from conscious participation in nonhuman animal exploitation through food and other choices for fifteen years, in July 2016 I received a crude reminder of the profound perversity of the system as it affects nonhuman females. The manager of a 'farm animal' sanctuary asked if I could help look after a very sick pig over a weekend as they would be busy organising an event, and the pig, Liza, required round-the-clock care. Liza, a pig rescued from a 'backyard farm', had been sick for close to two weeks, and as she was refusing food and water she was growing weaker by the day. Results of the veterinary examination were inconclusive as far as her sudden sickness was concerned, but showed that Liza was pregnant with around four weeks left before expected delivery. The pregnancy stressed her body further, there was little hope left for her and the piglets would most likely die with her; a caesarean was not an option as she was too weak for the procedure. In the sanctuary infirmary attempts were being made to syringe-feed and water her, and the bedding was changed whenever soiling occurred. This was going to be my job for that weekend, but on the morning when I was preparing to travel to the sanctuary, the manager called advising not to go to the trouble of making the trip as Liza would probably be dead by the time I arrived. For the past two days she had clenched her teeth very tightly and whatever food or water they managed to inject into her mouth would pour out unswallowed. The manager did, however, ask if I knew a healer who worked long-distance. I did not know healers, but the mention of healers (and ignoring the question of whether I had any right to try to keep her alive possibly against her will) reminded me of the power of psycho-bioregulation through physical proximity that I had been reading and writing about for several years. This helped me resolve to travel to the sanctuary as arranged. The (farfetched) proposition was to help her body re-establish some kind of balance by offering it my own healthy rhythms and neuroendocrine functioning to attune to or to take from it anything that may potentially benefit her own body/mind.

When I reached the sanctuary Liza was lying on her side motionless. I lay down behind her, our bodies touching, my chin resting gently on her forehead, and my arm wrapped around her belly, lightly so as to not put any pressure on her already heavily burdened interior. We lay there for the rest of the day, calm and quiet. Resting heart rates between adult humans and pigs are comparable with sixty to 100 beats per minute for adult humans and seventy to 120 for pigs,[85] and that may have worked to her advantage. There is a greater discrepancy in resting respiratory rates with twelve to sixteen breaths per minute for humans and thirty-two to fifty-eight breaths per minute for pigs, bringing pigs' rates closer to human babies between birth and six weeks of age with thirty to sixty breaths per minute. Having a continuous caring presence appeared beneficial for Liza: her breathing, slightly erratic at first, began to stabilise, and a few hours later when I tried to feed her she opened her mouth.

She survived another night, and to everyone's surprise, the following morning went into labour, delivering her first dead baby. She

---

85   Data for humans retrieved from NY State Department of Health, available at: https://on.ny.gov/2FbreT5; data for pigs retrieved from MSD Veterinary Manual available at: https://msdmnls.co/3iCxkJT

struggled all day and all of the following night. The contractions were weak and she had no strength to push. The veterinarian was called and injected her with oxytocin to induce contractions but to no avail. It is unclear what happened in our adult version of skin-to-skin care; however, our intimate proximity may have contributed to the release of endogenous oxytocin and this may have been the reason she went into premature labour in the first place. It took several hours between deliveries, and further inducement of contractions by mechanical means. This had to be performed slowly and carefully in consideration of her fragile state and exhausted body, with enough time between one act of human interference and another to allow her to rest. She cooperated as best she could, at last showing willingness to live. All of the twelve piglets were born dead, but Liza eventually recovered completely and now lives as a happy and healthy pig with her three children from her previous pregnancy.

Spending two days in such close and intimate proximity to a pig, nurturing her back to life, unavoidably triggers associations to other animals less fortunate than Liza, animals trapped on farms with no one to lend a caring hand when they are most vulnerable. On an industrial farm Liza would most likely have been prodded or kicked a few times looking for signs of life, had 'destroy' or similar wording painted down her back and eventually been taken to slaughter. Seen as mere breeding machines, or sources of feminised protein, as Carol Adams refers to dairy products and eggs, there is no space for considerations of the psychological and physical toll repeated pregnancies and subsequent removal of the children may take on these mothers. Unlike the Alto mothers, nonhuman animals would most likely not be able to rely on elaborate religious frameworks to help them cope with loss when the baby dies or is taken away from them. Feeling that they have no power to protect their babies would represent an additional burden on these mothers as would, in the case of forced removal, the fact that they do not know where the babies are being taken, and what will happen to them (and how).

Peer support may be available and may vary between different industrial settings. For instance, places where animals have more space to roam in relative freedom and the overall anxiety levels are lower may facilitate friendly relationships and offer more social support, increasing perhaps the chances of life as opposed to just survival, compared to heavily intensified operations where animals have to fight over personal space or even food and water. Anecdotal evidence, which is sometimes supported by video footage, shows that these nonhuman mothers will attempt to protect their children from humans. For example, when a new farmhand is first given the task of removing a calf from the mother on a dairy farm, the mother may attempt to fight back until the farmhand learns how to 'handle' the cow (for example, by hosing her away). The cow, realising her powerlessness yet again, or 'learned helplessness' as it has been called,[86] will not try it next time with this particular human, but may do so when the next new worker is hired, as a dairy worker once explained to me.

Attuned to their bodies, animals are able to detect changes, including those associated with pregnancy, which trigger various psychobiological needs, often ignored by the oppressive system of nonhuman animal use. As already mentioned, pigs, for instance, normally build nests for their babies prior to parturition, a phylogenetic imperative aimed at keeping the piglets warm and safe from predators. The extensive use of farrowing crates in industrial farming prevents this natural behaviour, substantially increasing the sows' stress levels and further jeopardising wellbeing more generally, including effects on parturition and lactation performance.[87] Trapped in a foreign environment with little or no self-determination and control over their own destiny and that of their children, these nonhuman mothers have certainly retained the propensity to love and care for their infants, but they are undoubtedly also aware of the reality of their circumstances; experience and observation have taught them that even if the children are born alive they will likely be stolen from them. Broken in spirit and resigned to their fate, or perhaps holding themselves together in a stoic fashion, or a combination of both, these mothers' grief may go unnoticed but just because it is not detected (who is paying attention anyway?) it does not mean that it is not there and running deep. 'Getting used to it' and adjusting one's behaviour to repeated separations does not necessarily make the loss less painful.

---

86   Maier and Seligman 2017.
87   See Yun and Valros 2015 for a comprehensive review.

Most nonhuman animals (and indeed most humans) do not have the privilege of living in the comfort and safety that affluent Western society currently offers to its privileged population, which probably includes many of the scholars and other commentators who share their opinions on grief in nonhuman animals publicly through writing and other forms of mass communication, influencing the public accordingly. Instead, nonhuman animals generally live in severely adverse conditions, either as victims of human exploitation or as victims of human greed leading to 'destruction of habitat' – a euphemistic phrase commonly used to describe these animals' potential or actual extermination. The expression of grief (not necessarily grief itself) could in fact be seen as a privilege, which many human and nonhuman animals might not be able to afford. Showing physical or other types of vulnerability in certain environments may increase danger: in the wild, it could attract predators, while animals on farms, when they slow down, are sent to slaughter. Further, prolonged grief with depression-like symptoms may dull our senses, making us less aware of our surroundings and hence less fit to cope with them. For instance, studies have demonstrated negative effects of depression on visual and olfactory perception.[88] Taken together, stoicism in relation to physical and emotional pain, along with the urgency to overcome acute grief in a hasty manner to avoid long-term, potentially detrimental outcomes, may manifest as a vital quality in many animals' communities, but as always, when it comes to grief, the results will depend substantially on the bereaved individual's psychological fabric and the nature of the lost relationship.

Weird and beyond

Stoic human societies are interesting for many reasons, including, and relevant to the current discussion, for their repression of grief expression. In the absence of a common language and relying on behavioural observations, one could easily (and erroneously) dismiss the existence of grief in these societies. They also provide yet another example of the diversity of grief expression/repression within humanity,

---

88   Bubl et al. 2010; Negoias et al. 2010.

which continues to be overlooked by those who would like to keep the imaginary divide between Us (humans) and Them (nonhumans) alive. We discussed the Alto community's stoic attitude towards infant death at some length. However, the Alto's dire conditions were relatively short lived. The repression of mourning, which developed in response to the immediate situation, ceased when the socio-economic conditions improved in the 1990s. It had been a societal response, a way of coping with the current reality; it did not penetrate the culture's psyche, it did not become a cultural norm, unlike in traditionally stoical cultures, like the Tigray and Bariba that we are about to meet. In fact, the Alto community, fortunately for its constituents one could argue, never developed as a culture per se.

Dag Ø. Nordanger conducted a study on coping with loss and bereavement among the stoic Tigray people of northern Ethiopia, following the Eritrean-Ethiopian war (1998–2000).[89] He found that grieving and crying are not socially condoned; they are believed to dry out the eyes and the joints, cause illness and bring misfortune. They do not bring the lost person back; they only distract from work that is needed to support the remaining family, and they provoke God for not trusting him and for craving for his property (the deceased). In an economically and climatically dire environment where everyone's labour is critical to support the family, where people cannot get sick leave to work through their grief, Nordanger notes, it is important to focus on the present; expressions of grief are thus avoided; the coping mechanism is 'forgetting', and, as in the case of the Alto community, the safety of religious imagination. This of course does not mean that these people are not grieving inside, as Nordanger also recognises, in agreement with Thera Mjaaland whom he cites:

> If I had come for just a short visit without knowing anything about what the people have been living through, I would have thought they had always been happy and gay; poor, but seemingly chatty, cheerful and laughing most of the time . . . There are no obvious signs of their past suffering in how people

---

89   Nordanger 2007.

behave in their daily lives, and in what they choose to talk about; unless I ask them in private.[90]

The Bariba people of Benin, one of the epitomes of stoicism, also abstain from expressing emotions publicly. The subject of pain in the Bariba community, as anthropologist Carolyn Sargent, who conducted a study on this phenomenon, notes, 'evokes a cognitive map of honor and shame rather than a discussion of pain per se.'[91] Courage is central to their identity and showing pain indicates cowardice. Cowardice is the essence of shame, and between shame and death the Bariba will choose death, as one of their proverbs clearly illustrates: 'sekuru ka go go buram bo – between death and shame, death has the greater beauty.'[92] This applies to both physical and emotional pain; the entire culture is oriented towards building courage, and children are taught from an early age to suppress verbal or behavioural displays of fear and pain. At the age of six, a child is expected to be able to control emotional responses and understand the concept of shame, and thus respond to painful stimuli accordingly. Initiation rituals, which include bodily mutilation (for example, circumcision and clitoridectomy), offer an unparalleled opportunity to demonstrate one's courage. Similarly to physical suffering, emotional pain is also to be disguised: a grieving person, for instance, may weep in private, but should not express their feelings publicly. While the Bariba appear to be more tolerant of female lapses and more demanding towards men, in other stoic societies the opposite may be true, if the following instruction reportedly conveyed by Chaga grandmothers to their granddaughters can serve as an indication: '[I]t is a man's nature to groan like a goat when in distress, but women suffer silently as do sheep.'[93] Goats in physical pain indeed vocalise more readily compared to sheep and cows,[94] but this does not mean that the latter suffer less.

---

90   Mjaaland 2004, 43; cited in Nordanger 2007, 558.
91   Sargent 1984, 1299.
92   Sargent 1984, 1299–1300.
93   Raum 1940, 84, reported in Sargent 1984, 1303.
94   National Research Council 2009, 61.

Stoicism goes a step further in Bali where the bereaved are not only expected to show composure but conventional mourning practices require them to appear cheerful.[95] The Chocorvinos in the Peruvian Andes, on the other hand, have no proscriptions against expressing grief publicly; their funerals, as I. Neil Stevenson notes, 'are accompanied by loud and dramatic expressions of grief which are encouraged by drinking and chewing cocoa leaves among the mourners.'[96]

Other human societies around the world approach grief and mourning in ways that would be considered unacceptable in the West and may not necessarily be recognised as grief and mourning by a Western observer: this is critical to keep in mind in considerations of nonhuman animal grief and grief expression. For instance, the funeral of a young, twenty-eight-year-old Asante man was 'a pandemonium of vehement crying, shouting, tooting horns, men in women's dresses and fighting around the grave', reports medical anthropologist Sjaak van der Geest, with a woman 'waving condoms indicating the incompleteness of the young man's life'.[97] If, on the other hand, someone commits suicide, the ritual changes dramatically. Dying during childbirth (when the baby also dies) is considered to be the worst kind of suicide, an act of cowardice, because instead of carrying out the most honourable duty, which is that of bringing a child to life, as they see it, the woman kills the child. In this case, traditionally, the body would be dragged into the woods, where other pregnant women would spit on it and verbally abuse it while pointing plantain leaves at it.[98] Nowadays with this practice ceased, the women are given a minimalistic funeral.

Moving south-eastward, specifically to the Dedza-Angonia borderland (between Malawi and Mozambique), we find further mourning and funeral practices, which differ substantially from the conventional Western method and which Harri Englund describes in some detail.[99] These include women (relatives and other villagers) remaining in the house with the deceased, weeping and wailing

---

95   Schaefer 1999, 14.
96   Stevenson 1977, 304.
97   van der Geest 2004, 905.
98   van der Geest 2004, 904, citing Bartle 1977, 378.
99   Englund 1998.

continuously, while the men, outside where the mood is strikingly different, make jokes largely revolving around sexuality, including obscene remarks on wives, humour and laughter being the norm. The mourners may also create a caricature of the deceased by wearing a characteristic item of clothing, and imitating and exaggerating the deceased's manners. The purpose of this imitation, the villagers claim, is not to offend the deceased; rather, it is the last exchange between the deceased and the living.

Sexually themed rituals occur in other societies, too. Huntington and Metcalf point out that death is not only about the deceased and the transition into an afterlife, it is also about life and the living, the life of the deceased, the life she or he has left behind, including progeny:

> The continuity of the living is a more palpable reality than the continuity of the dead. Consequently, it is common for life values of sexuality and fertility to dominate the symbolism of funerals.[100]

The funeral ritual in the Bara community of south-central Madagascar, described by Huntington and Metcalf, is a case in point.[101] There is strict separation of women and men during the day with women remaining with the corpse in 'the house of many tears' (so named because the vigil includes ritualistic outbursts of loud weeping), while men stay at a separate house where they also keep a vigil and tailor the logistics of burial. This segregation is interrupted at night when both genders come together and engage in wild drinking, dancing and singing with a strong sexual component, which reflects also in the funeral songs sung by the females, for instance:

> *now hide it*
> *now hide it, boys*
> *now hide it because there is a death*
> *together let us copulate*
> *together let us copulate, boys*

---

100 Huntington and Metcalf 1979, 93.
101 Huntington and Metcalf 1979, 102–18.

…
'brroo' flies the quail
to perch on a bump of a sakoa tree
the eye wants to copulate
the eye wants to ejaculate …

where 'brroo' is a common expression for ejaculation and 'eye' refers to any centre, hole, vortex, including the vagina.

Other rituals foreign to the West include, for instance, lying down beside the corpse, smoking cigarettes and offering cigarettes and food to the corpse, as the Berawan of Borneo do; widows may have to rub the decomposing matter from the husband's corpse onto their own bodies.[102] If such interaction with the corpse, particularly the offering of food, which is not an uncommon ritual in human societies, is observed in a nonhuman animal, humans are quick to declare the animal incapable of understanding death, while in fact the animal may simply be trying to ascertain whether the individual is truly dead, and in the case of a close other the animal's interaction with the corpse is most probably part of grieving. Avoiding touching the corpse as in the case of the dogs discussed earlier also finds parallels in the human world, such as the biblical prohibition in Numbers 19:11: 'The one who touches the corpse of any person shall be unclean for seven days.'

The final disposal of the corpse also varies across human cultures. Some bury their dead, others do not. Some bury the body intact, others opt for cremation. Rotting in some cultures, such as the Berawan, is considered a positive phenomenon (they even store the corpses in the same kind of barrels used for rice wine to then drain the liquid from the decay and save the bones), as it releases the dead person from the bonds of flesh.[103] In Cambodia cremation is favoured over burial in the belief that slowly decaying bodies give rise to evil spirits.[104]

Those who do not bury the dead have also developed various methods of treatment and disposal of the corpse. The Sioux, for

102 Huntington and Metcalf 1978, 64, 74–5.
103 Huntington and Metcalf 1978, 56, 64.
104 Boehnlein 1987, 767.

instance, have traditionally wrapped the corpse in a buffalo garment and placed it in a tree. There the corpse would provide sustenance to birds and other small animals; later the corpse would drop onto the ground feeding larger animals and eventually regenerating the earth itself, which in return will grow grass for the buffalo, whom the Sioux will then kill, closing the 'circle of life', as they traditionally see it.[105] The Maasai of Kenya and northern Tanzania also leave corpses out for other animals to eat, but unlike the Sioux, they believe that corpses are harmful to the soil.[106] In a similar vein, the Yoruba people of Nigeria may throw dead babies and stillborn into the bush without burying them out of the belief that burying such a child 'would offend the shrines of fertility'.[107] The Tibetans may also choose to avoid ground burial and opt for so-called 'sky burial', which consists of cutting up the dead body and feeding it to the birds, on a mountaintop.[108]

Cannibalism is also not unknown in human cultures as a corpse-disposal ritual. The Wari' people, introduced in Chapter 1, traditionally consumed the bodies of their deceased in-laws as part of the mourning ritual. In light of this cultural phenomenon, what Watson and Matsuzawa, referring to nonhuman primate mothers eating parts of the dead infant that they are carrying around lovingly, describe as the 'juxtaposition of care and cannibalism'[109] may be less puzzling than the authors imagine.

* * *

The variety of responses to death and grief (the meanings of which we learn through our shared verbal capacities) within the human world demands caution in attempting to evaluate the presence or absence of grief in other animals based on behavioural observations. The absence of outer expressions of grief as in the case of the stoic Bariba people is no indication of the inner subjective state of the bereaved, just as

---

105 Schaefer 1999, 8.
106 Reported in Arvin and Arvin 2010, 258.
107 Berman 2001, 9.
108 Martin 1996.
109 Watson and Matsuzawa 2018, 3.

the continuous wailing of a Mozambican female neighbour induced by cultural etiquette to participate in the mourning ritual need not indicate her feelings towards the deceased.

The importance, in the animal world, of being able to distinguish between a living organism and a dead one along with animals' frequent encounters with both, at least in the wild, strongly supports the hypothesis that nonhuman animals 'understand' death as permanent cessation of agency and physical non-returnability. The psycho-neurobiological comparability between humans and other animals along with our shared capacity and propensity for bond formation suggests that when loss of a close individual occurs, human and nonhuman subjective experiences are also comparable. Clearly, many human cultures have developed elaborate rituals surrounding death. Rituals have been shown to substantially mitigate the feeling of grief and help re-establish a sense of control, which may be needed after the internal world has been shattered following a loss.[110] Explicit death rituals have been observed in some nonhuman animal species, as noted at the beginning of this chapter, while many rituals in many other animals may have for various reasons gone unnoticed: they may have been too subtle for the human observer to catch, too different from what the observers would consider a mourning ritual for them to have recognised it as such, or simply never seen because no human witness was in the 'right' place at the 'right' time. This is to speak of *visible* rituals and expressions only, regardless of what *invisible* rituals may be occurring in the animals' (human or not) *milieu intérieur*, while recognising that nonhuman animals do not appear as afflicted by *things* as humans.[111]

The principal difference between human and nonhuman animals in relation to death appears to be far more complex and covert. When nonhuman animals die they do not usually take other animals with them. Humans, on the other hand, have tended to do so: when humans die, the grief is shared across species – not because other animals willingly offer themselves as sacrifices but because humans force the sacrifice onto them. This can take the form of a direct, explicit sacrifice

---

110 Norton and Gino 2014.
111 David Brooks, personal conversation, 2018.

as in the case of 'cattle' in the Bara funerals, whom the humans force to wrestle at the burial site. This 'sport', as Huntington and Metcalf call it,

> of stampeding the herd around and around resembles a bovine version of the funeral dances. When describing the event, the Bara boys always emphasize the snorting, panting, and bucking of the cattle as signs of intense vitality.[112]

This may be hard to deny – the vitality, their willingness to live, their desire to not be harmed, which is converted into a more palatable account of the cow's mind, reflected in fragments of folk fiction, in which the cow speaks in a clear act of anthropomorphism:

> When I die do not bury me in a tomb but your stomachs shall be my tomb. My head you shall not eat. Bury it in the earth. After one week corn sprouts and rice, manioc, and sweet potatoes. And the herd too shall give you life.[113]

It can also take the form of an indirect sacrificial act, such as ordering the flesh and feminised protein of various animals, Liza's sister, or perhaps her nephews, from a local deli for a more mundane, contemporary Western funeral and mourning ritual, all blood cleaned away or disguised in some way. In either case, nonhuman animals are taken from their communities (or prevented from having a normative community in the first place), and subjected to a painful death as part of humans' own mourning rituals while the feelings of the victims themselves and those who care about them are either contorted or plainly ignored. Modes of death and grief expression may differ across species; the same, however, could not be stated with any certitude for the feeling of grief itself.

Departing from the discussion of death and grief, we will now explore animals' implicit celebrations of life beyond intersubjective attachment and love. We, animals, are immersed in an inescapable

---

112 Huntington and Metcalf 1978, 112.
113 Huntington and Metcalf 1997, 112–3, citing Faublée 1947, 381.

dance of relationality, captured succinctly by Christina Rossetti in the
following passage from her poem 'To what purpose is this waste?'

> *And other eyes than ours*
> *were made to look on flowers,*
> *eyes of small birds and insects small:*
> *the deep sun-blushing rose*
> *round which the prickles close*
> *opens her bosom to them all.*
> *The tiniest living thing*
> *that soars on feathered wing,*
> *or crawls among the long grass out of sight,*
> *has just as good a right*
> *to its appointed portion of delight*
> *as any King.*[114]

---

114 Rossetti 1896.

4
Spiritual animal

Wouldn't it be funny if the ability to immerse oneself in the flow of life, that is, to live for a moment fully in the waves of the present – a state thought to determine animal modes of awareness as inferior to human consciousness – is in fact not the lowest stage of consciousness but the highest?
— Cynthia Willett, *Interspecies Ethics*, 2014[1]

No Beast is there without glimmer of
         infinity;
no eye so vile nor abject that brushes not
against lightning from on high, now
         tender, now fierce.
— Victor Hugo, *La Légende des siècles*, 1859[2]

Nonhuman animal brains share human propensity to understand and categorise events – make some sense of them within the broader context of their socio-natural environment and their own lives. Do

---

1   Willett 2014, 130.
2   Cited in Kristeva 1982, 1.

other animals also have the capacity, or even the interest, as some humans do, to find meanings of life and death in some abstract world 'beyond', constructed through religious imagination?

If, for a moment, for lack of evidence in either direction, we assume that they do not, does this preclude them from meaningfully engaging in relationality with their environment, including with intangible agencies in it, 'transporting ordinary life into a dream of cosmic peace',[3] which we could label as *spiritual*? If the answer is no – if we allow the possibility of nonhuman animals as spiritual subjects, as agents of embodied spiritual exchanges – how much more tragic captivity is than previously imagined, particularly in its forms of intense confinement where animals are deprived of stimuli that would in species-specific ways trigger such engagement and the appreciation and celebration of life at this deeply implicit but experientially meaningful level.

In Western discourse, the consideration of nonhuman animals as spiritual and/or religious subjects has, like other aspects of their being, largely been tainted by assumptions of an ontological determinism, which denies nonhuman animals agentic properties in the perception and formation of the world they inhabit. When not completely ignored or instrumentalised for human conceptual or more mundane benefits (for example, as metaphors[4] or food), nonhuman animals' existence may be perceived as purer and their experience of the world as more immediate. They may be seen, to borrow Bataille's wording, as being 'in the world like water in water',[5] or even, as Rilke suggests in 'The First Elegy', 'aware that we [humans] are not really at home in our interpreted world'.[6] Nevertheless, to a greater or lesser extent, nonhuman animals' generally remain conceived as instinctual automatons, their experience

---

3   Willett 2013, 187.
4   Nonhuman animals are often used as metaphors to describe humans'
     shortcomings (e.g. chicken for cowards, sheep for followers). The
     characteristics attributed to various species tend to be extracted from the
     context and may have little to do with how these animals truly are in real life;
     nevertheless they do impact upon human perceptions of them and as a
     consequence also upon the lives of real flesh-and-blood animal individuals
     and species (e.g. Merskin 2018).
5   Bataille 1989, 23.
6   Rilke 2009 [1923].

both qualitatively and quantitatively different from that of humans. Consequently, speaking of nonhuman animals as spiritual subjects/ agents continues to draw accusations of anthropomorphism.[7] The principal reason for such accusations appears to be the emphasis on the cognitive, interpretative domain in discussions of spirituality. Commentators tend to assume that a spiritual experience only acquires actual experiential meaning, is only felt to be experientially significant, when it is filtered through a declarative meaning-making process and interpreted and contextualised, whether through religious imagination or some other non-denominational cognitive framework. Most humans do not consider nonhuman animals' cognitive capacities sophisticated enough for this kind of abstract meaning-making and as a result they dismiss the possibility of spirituality in nonhuman animals.

Simultaneously, concerned commentators and practitioners also tend to agree that it is this very aspect of the self – the cognitive, declarative, reflective self – that needs to be relinquished or *transcended* in order to enable a holistic embodied communication underlying spiritual experience. However, the notion of transcending the self – self-transcendence – is problematic for at least two reasons. First, it may mislead us into thinking that a human-grade reflective self is a prerequisite on the way to spiritual experience. This may close the circle of human exclusivism in spiritual matters, foreclose the possibility of nonhuman animal spirituality according to the following (invalid) formula: the human reflective self needs to exist in order to be transcended; transcending this self enables a spiritual experience; this experience must then be interpreted through the declarative apparatus to become truly meaningful to the reflective self. Second, 'self-transcendence' is a negative descriptor, telling what spirituality is *not* rather than what it is or may be. We will not find it in the declarative self; we need to transcend this self but where do we go from there?

In contrast to the above conceptualisations, I suggest that the sense of oneness, of merging, of connection that researchers and religious practitioners often cite in considerations of spirituality, materialises (in the sense of becoming experientially graspable) through a process of

---

7   e.g. Fisher 2005.

reaching into the realm of the implicit, experiential self. Conceptualising this process as 'self-transcendence' attributes to the reflective self a primacy in spiritual experiences that the reflective self does not hold. Simultaneously it obscures the role of the experiential self – the level of consciousness that informs the organism that it is alive and feeling, the level that a living organism thus cannot transcend. It is not the reflective self but the experiential self that is fundamental for spiritual experience, for it is the experiential self that communicates with intangible agencies during a spiritual exchange, and this communication, this experience, is felt by the organism without the reflective self's interference and post-processing. Critically, both human and non-human animals have equal access to the experiential self.[8]

Essentially, two distinct processes emerge in relation to the self, both of which are normative for the animal (human inclusive) brain and have their own significant functions. On one level there is the implicit experiential self, which is intrinsically relational and in constant communication intra-organismically (within the organism itself) and with the outside environment. This level of consciousness, as cumulative scientific evidence suggests and as discussed earlier in this volume, is already felt, experientially meaningful, without needing to be 'converted' into meaning by some interpretative cognitive functions. The interpretative level, on the other hand, encompasses processes that are reductionist in nature, translating holistic experience into a

---

8   I need to clarify here that this is not a discussion on the more commonly known and discussed differences between immanent and transcendent perspectives/traditions, both of which engage the experiential and reflective selves. What I wish to emphasise is that regardless of the tradition, a spiritual experience takes place on the level of the experiential self. However, as noted later in this chapter, certain perspectives may facilitate the fluidity of engagement of the experiential self to a greater extent than others. For instance, someone who is already fully immersed in the flow of life of the community of subjects (including nonhumans) on Earth may attend to the relational potentialities that may lead to a spiritual experience differently compared to someone with a dualist and anthropocentric orientation. Nevertheless, even in transcendentally grounded traditions, argues Latour (2010, 105–6), the focus is/should not be on the 'far away' but the close, the present and, conversely, immanent traditions are not devoid of transcendental features either (e.g. Crosby identifies several in his *Religion of Nature* (2003)).

depleted yet functional reality. This level enables religious imagination and other forms of contextualisation, but it does not generate spiritual experience per se. However, the two levels tend to be conflated in discussions of spirituality. This ultimately leads to denying that nonhuman animals possess capacities for spiritual experiences.

As academic interest in nonhuman animals grows, prejudicial barriers that the Western mind has erected against them are progressively collapsing. James B. Harrod and Donovan O. Schaefer, for instance, offer much food for thought in their deliberations of nonhuman animal religion.[9] They also introduce useful language for challenging implicit concepts of earlier types of discourse. Borrowing some of that language,[10] and in agreement with Aaron S. Gross's emphasis on relations and agencies,[11] I invite the reader to envision spiritual engagement as animal bodies' affective *dancing with animacy*. From conception to death, the organism is in constant communication with the environment. The environment acquires animacy: it becomes animated and endowed with agency by its capacity to speak and respond to the organism on an experiential level. As discussed earlier, this happens well before (and continues after) the organism develops the ability for any kind of reflective consciousness, leaving the organism essentially primed for affective communication with the rest of the world.

In other words, we can conceptualise the spiritual experience as the engagement of the individual animal (human and other) with agency as a quality of entities/phenomena that materialises by way of producing an effect on the animal's experiential consciousness, drawing the animal into the relational dance. Such agentic potential thus is not confined ontologically but it belongs to all life forms and perceived manifestations. This exchange between the organism and animacy from the environment resembles dancing in the sense that the interaction of

---

9   Harrod 2011; 2014; 2016; Schaefer 2012; 2015.
10 Harrod (2011, 344) defines religious behaviour as 'ritualization of empathic intimacy in relation to animacy', while Schaefer (2015, 192) sees religion as 'a dance of relating ... a compulsory, affective web fusing bodies to worlds'. See also Johnston 2017.
11 Gross 2015.

parts gives rise to a whole, and this whole can only be sustained while the parts continue to interact in synergy.

During the dance the lines between the self and the nonself become blurred, producing a feeling of merging and oneness. However, this oneness, much like when two people dance or make love, should not be understood as a dissipation of the self; rather, it is a connection with a sense of presence that arises in this affective communication among the various cues within the organism and those reaching it from the exterior. This view is not too dissimilar to Schaefer's conceptualisations of religion.[12] However, in my view religion is a broader concept that includes implicit and explicit elements for human and nonhuman animals alike, given that the latter are also endowed with interpretative and decision-making capacities. In contrast, the focus of this chapter is the implicit dimension that exists outside interpretative frameworks: spirituality, rather than religiosity, which manifests as a propensity of animals' intrinsically relational, non-reflective, experiential, embodied consciousness.

Schematically, this relational dance could be presented as follows:

12　Schaefer proposes to move beyond the logocentric, human exclusivist view and explore religion from an evolutionary perspective, stressing the importance of affect, which humans share with other animals. Schaefer theorises religion, as partly already seen in a previous footnote, as 'a dance of relating, a compulsory, affective web fusing bodies to worlds [...] a cycling of semistable bodily forms' emerging differently in different bodies (p. 192). Quoting LeMothe, Schaefer agrees that religion 'is not what a person believes, nor what he has or does per se. Religion rather exists *in the moment of its performances as a kind of doing that embodies a person in relation to a sense of the world*' (p. 191). According to Schaefer, faith and belief, for example, are elements of the human religious dance, while other bodies, other animal species, with their distinct evolutionary histories informing distinct biophilic relations and related affective triggers, engage with the world differently, which manifests in different forms of religious dance.

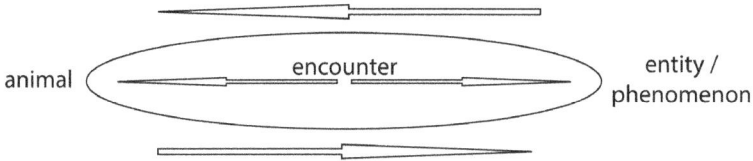

animal — encounter — entity / phenomenon

Encounter between animal and entity/phenomenon.
Entity/phenomenon speaks to animal = animal is affected.
This confers agency to entity/phenomenon. Agencies (of animal and entity/phenomenon) interact.
This exchange/relational engagement between agencies = spirituality.

Note that the entity/phenomenon can possess agency on its own, irrespective of the encounter, but from the animal's perspective it is only recognised as meaningful if it affects the animal.

Let us consider an example of spiritual vs non-spiritual engagement: the taste of madeleine cakes can produce a spiritual experience if it speaks directly to the experiential consciousness, as it does in Proust, discussed later on; whereas a non-spiritual experience of madeleines is produced when, for instance, you take the cakes out of the oven, try them and feel thrilled that they are literally melting in your mouth (unlike last time you made them), and won't the guests love them! Of course there could be confluence of both.

The emergence of a spiritual experience is facilitated by the non-reflective experiential self and the experience is modulated by organic processes that cannot be reduced to particularities of human brain organisation. A significant intellectual barrier to this argument includes the myth of nonhuman animals and their societies as instinctive automata – a myth, as already indicated, that persists despite cumulative evidence to the contrary. The reductionist yet necessary processes of interpreting and categorising experience, which in some humans may give rise to religious imagination, exist also in other animals. This enables group structure, social norms, and moral codes, and, generally, facilitates individuals' navigation in their socio-natural environment. Furthermore, recent scientific studies in the area of self-relatedness help to challenge previous Western dualist assumptions and consider the way the self is constructed and lived, problematising the concept of self-transcendence and allowing the more accurate notion of self-extension to become prioritised in further considerations of spirituality.

Religious imagination vs spiritual engagement

The elusive nature of religion and religiosity becomes perhaps most evident when attempts at defining the phenomena are made. Definitions range from narrower views of religion as a set of beliefs in so-called supernatural beings and rituals associated with it, to broad and possibly overgeneralised notions whereby just about any act signalling some form of ethical orientation or contemplative, meditative disposition is taken to indicate religiosity. The disputed adequacy of the term in comparative studies and the lack of an agreed definition have even led to suggestions that the term should be dropped entirely. Daniel Dubuisson, for example, proposes to replace it with the concept of 'cosmographic formations', which would cover various aspects of the worlds as constructed by humans.[13] As a term born out of the Christian tradition, Dubuisson finds attempts at reinterpreting and redefining religion in order to strip it of its ideological heritage and render it suitable for cross-cultural application fruitless. Rejecting the idea that humans are inherently religious, 'cosmographic formations', in Dubuisson's view, would represent a universally applicable and scientifically valid framework for the study of humanity, inclusive of materialist, atheist and agnostic formations.

Since the publication of his French volume in 1998, substantial work has been done to address the Eurocentricity of the term and to expand the field of religion studies to accommodate the existing human cultural (religious?) diversity as well as that of other animals. Nevertheless, in my view, there is much to appreciate about the proposed terminological shift, including greater cross-species applicability in the exploration of phenomena that many humans still

---

13   Dubuisson 2003 [1998]; for a compilation of reviews and Dubuisson's responses to them see Engler and Miller 2006.

guard as a human-only domain, and space to avoid attributions of religiosity to formations such as Western Atheism, which emerged specifically to contrast the existing oppressive religious powers. Of course, all formations are borne out of the human mental apparatus, making parallels rather unavoidable, yet 'religion' remains a value-laden term and sensitivities to the use of the term will persist. It is unlikely that a concept, which has contributed substantially, for better or for worse, to the formation of the Western mindset (theist or not) from which relevant academic disciplines (for example, anthropology and religion studies) emerged, will subside, and perhaps it does not need to. However, it is to be hoped that the complexity with which the concept of religion is endowed within its own meta-discipline spreads to other disciplines and to the general public, both of which seem to have retained a rather orthodox view of religion and religiosity.

Like 'religion', the term 'spirituality' is also replete with ideological connotations, stemming in part from attempts to identify expressions and manifestations that could serve as criteria to describe religion. More often than not, Del Rio and White note, the terms are used interchangeably regardless of whether the various authors advocate for separation or coalescence of the two phenomena.[14] Concerned with the widespread obfuscation of the two notions – coalesced also in the *Diagnostic and Statistical Manual of Mental Disorders* – Del Rio and White attempt an elaboration of the distinction, highlighting its merits in particular for health-care professionals and their clinical approaches, which would obviously benefit from conceptual clarity. The article endorses a rather traditional understanding of religion as a set of (learnt, adopted) beliefs and rituals entailing community, whereas spirituality is seen as innate to humans, entailing 'human individuality and making sense of one's life.'[15]

Even though most authors seem to recognise the centrality of relational intimacy for spirituality, the articulation of such appears hindered by the reluctance to reach beyond the cognitive paradigm. In their attempt to separate religion from spirituality Del Rio and White begin to inadvertently attribute properties (meaning of life, how one is to live, etc.) to spirituality that may be better suited to the realm

---

14   Del Rio and White 2012, 124.
15   Del Rio and White 2012, 131.

of philosophy. The authors try to rectify the inadequacy of these descriptors by adding so-called transcendence to the mix. However, regrettably, they fail to define transcendence, and they do so intentionally. They fear that addressing the question of 'what is transcendent' at this early stage in their exploration of spirituality would lead to further conflation of religion and spirituality. Instead the authors leave us with a definition of spirituality constituted of philosophical questions with an added ('transcendent') component which, if left undefined, manifests substantial religious connotations.[16]

Others may accentuate the importance of 'the sacred' for spirituality. For Hill and colleagues, for instance, the criterion for spirituality is 'the feelings, thoughts, experiences, and behaviors that arise from a search for the sacred', whereby the sacred 'refers to a divine being, divine object, Ultimate Reality, or Ultimate Truth as perceived by the individual'.[17] Although the wording tends to imply the need for cognitive elaboration, the teleological nature of these and similar definitions suggests that the authors may indeed be in agreement with the view of spirituality advanced in the present work: ultimately spirituality is not the *quest* for meaning, 'transcendence', 'the sacred', as it is often proposed to be; rather, spirituality is the *encounter* with it/ them. Hill and the co-authors, in fact, specifying what they mean by the term 'search' in their criterion for spirituality, allow the possibility of 'search' signifying *maintaining* the sacred along with identifying, articulating or transforming it.

It is perhaps the millennia-long conditioning with 'reason' and its attempts to supplant emotion, and with theological abstractions accentuating the greatness of divinity compared to the humbleness of the created that, in the West, has generated the current widespread sense of impotency to feel and to conceive the relational dance with animacy directly rather than through cognitive labyrinths. Another, and more recent, phenomenon possibly hindering accuracy of conceptualisations of spirituality is the new stream of life-philosophy – the so-called 'spiritual but not religious'[18] people – which emerged

---

16   Del Rio and White 2012, 138.
17   Hill et al. 2000, 66.
18   Erlandson 2000.

following increasing detachment from organised religious communities in Westernised societies. Open to, and sometimes actively seeking, spiritual experience, these individuals and groups, despite their diversified lifestyles and ideas of the sacred, in essence represent another cosmographic formation. Whereas, the extent to which an individual may be considered spiritual – regardless of our ideological background – may better be measured by an individual's capacity for and recurrence of engaging in the dance with animacy and its inherent self-extension, instead of by their verbal commitments in that direction.

It is unclear to what extent the seemingly intense preoccupation with the nature, meaning and purpose of life is indeed inherent to the condition of being human, or whether such preoccupation may have been inflated and transferred on to us by traditions that have led themselves to the need to articulate sophisticated ontological and existential frameworks following conceptual extraction of the human from the rest of the living world. The question of *how* we are to live in the world and with others, on the other hand, *is* intrinsic to humans as well as to other animals.

The codes of conduct incorporated in religious teachings reflect the normative tendency for cognitive closure within a broader context of questions related to communitarian living, delineating acceptable and unacceptable conduct and providing structure. This would increase predictability and therefore a greater sense of safety, and also enable smoother inter- and intra-communitarian relations. The presence of such codes in societies does not necessarily indicate religiosity unless we arbitrarily choose to define them as such (and then, in turn, re-use them as indicators of 'religiosity' in the existing definitional vortex). Neither does the presence of such codes in religious teachings reflect the 'intrinsic ethicality' of religions, as some would have it. Apart from prosocial behaviour, religious teachings, and societal conventions more broadly, may to various extents promote anti-social behaviour (for example, gender inequality or homophobia), and prescribe or proscribe behaviours that are ideologically exacerbated but, in fact, ethically neutral. For instance, I concur with Broom when he points out that laws regarding sexuality are generally grossly overstated in some religious contexts.[19] Working on

---

19   Broom 2003, 224.

the premise that there is consensus among the parties directly involved, there is nothing intrinsically ethical or unethical about practices such as monogamy or heterosexuality, but some religious traditions may prescribe one or both as compulsory conduct and hasten to accuse those who do not comply (including nonhuman animals) of immorality or lack of morality. Non-violence, reciprocity, kindness, on the other hand, in my view belong to a different relational category because the non-observance generally translates into direct violence against another. Other animals are also perfectly capable of moral consideration, although admittedly, unlike many humans, they tend to show it in *facta non verba* – in deed not in word. Despite the commonly held belief to the contrary, convivial relations appear to be far more common among animals than aggressive behaviours, as discussed before. 'They are at pains to avoid conflict whenever they can,' emphasises de Waal.[20]

In essence, just like human lives, the lives of nonhuman animals are replete with evaluations, interpretations and decision-making in relation to both the physical and social environment they are embedded in. This process of meaning-making, which is not in itself spiritual, can – and perhaps often does – include a spiritual dimension insofar as it allows space for opening, for an intimate encounter of the experiential self with animacy, which may subsequently inform cognitive closure and how we translate the intimate experience into everyday living.

As discussed at some length in Chapter 1, animal brains develop mental models that filter the raw data animals are exposed to in everyday life into a more manageable whole. This is a reductionist but normative function of the animal brain, helping all animals (including humans) make some sense of the world by structuralising and compartmentalising experiences and phenomena and finding relations and logical connections among various objects and actions. This process unavoidably depletes the holistic embodied experience but also takes away the potentially stressful feeling of randomness by making the world more predictable, enabling easier navigation and a greater sense of control and safety. It is this reductionist process that enables religion, among other things. Already in 1980 Stewart Guthrie, setting the foundations for what would later develop into the cognitive science

---

20   de Waal 2013, 227. See also Bekoff and Pierce 2009.

of religion, noted that 'there is no fundamental disjunction between religious and nonreligious modes of thought, despite their difference in content' and that 'religious and nonreligious belief and behavior can be best understood as closely related variants of a single human cognitive process'.[21] The more holistic right hemisphere needs the left to unpack experience. Indeed, without such unpacking and concept formation, according to McGilchrist, we would not have religion (its more technical components, not the affective ones), 'only' awe.[22] However, he adds, without further integration through the right hemisphere, all one is left with is theology.

Addressing awe specifically and building on prior work on uncertainty-tolerance and cognitive closure, Valdesolo and Graham found that the uncertainty and, generally, the negative feelings that may emerge at an awe-inspiring encounter as a consequence of the inability to assimilate novel information into existing mental schemas can be attenuated by attributing agentic properties to the phenomenon or to a force behind it (for example, God or karma).[23] However, people exhibit individual differences in the need for cognitive closure,[24] affecting the level of dispositional awe-proneness[25] and making some people more prone to awe than others. Awe-prone people show a higher uncertainty tolerance as well as increased comfort and flexibility in adjusting existing mental structures to accommodate new experience.

Exposure to dogmatic interpretations, such as may be the case in people coming from certain types of religious backgrounds, contributes to the shaping of the mental templates accordingly and is likely to influence the way people process an awe-inspiring phenomenon. For instance, the 'transcendent' experience of Buddhist monks during meditation and Franciscan nuns during prayer was reported by the study participants as connection with 'the void' and with God respectively.[26] In a similar vein, near-death experiences (NDE) and

---

21 Guthrie 1980, 181.
22 McGilchrist 2009, 199.
23 Valdesolo and Graham 2014.
24 Webster and Kruglanski 1994.
25 Shiota, Keltner, and Mossman 2007.
26 Johnstone et al. 2012, 280.

their post-experiential interpretations are also prone to cultural bias, as noted by Nelson, and by Morse who examined NDE in children and reported children seeing Jesus (who looked like Santa Claus) or secular scenes such as teachers and school friends, depending on their religious background or lack thereof.[27]

Kevin Nelson elaborates on the question of NDE, and spirituality more broadly, from a neurologist's perspective. He bases his considerations of spiritual matters largely on human experience following medical episodes (rather than on intrinsic relationality) but his conclusions are compelling. For instance, he attributes the 'losing' of the explicit self to a 'contraction of explicit self and an expansion of implicit self'.[28] While he exhibits a certain level of human exceptionalism in his presentation of animal consciousness and selfhood,[29] he nevertheless maintains that language is not a requirement for spiritual and mystical experiences, allowing the possibility of such experiences in other animals.[30] Discussing specifically the various aspects of the NDE (white light at the end of the tunnel, out-of-body experience, bliss, and related features), he suggests that they are likely a result of neurophysiological processes (for example, low blood flow to the eyes, REM sleep, and the brain's reward system) that humans share with other animals.

Returning to the interpretative domain, there is no reason to believe that the mental template employed in the evaluation of the

---

27 Nelson 2011, 104, 112; Morse 1994, 69–70.
28 Nelson 2011, 63.
29 e.g. on p. 54 Nelson dismisses the possibility of cockroaches having consciousness, citing, unconvincingly, human (*reflective* – by implication) consciousness and its unique content as some sort of justification/explanation for his scepticism; while on p. 63 Nelson acknowledges the existence of an implicit and explicit self, admitting that the borders between the two are not strictly defined; nevertheless the choice of examples to illustrate the difference between the two can be highly misleading for the reader: namely, Nelson takes an enormous leap from genetic coding, representing the implicit self, to yet another instance of human reflective consciousness (Nelson's example: you become reflectively and experientially conscious of your right foot when I draw attention to it), potentially giving the impression that anything between these two extremes is irrelevant or non-existent.
30 Nelson 2011, 254.

phenomenon in question determines how strong or meaningful the experience itself is for an individual. For some people the encounter with an awe-inspiring phenomenon may strengthen the love they feel for the mental construct they call God, but filtering the experience through a logocentric, God-fashioned template does not necessarily make the experience more meaningful compared to absorbing the experience without such cognitive closure. On the contrary, the promptness to assimilate the experience within established patterns may provide false security and a sense of mental ownership of the event, reducing some of the intensity.

When Christopher L. Fisher dismissed Jane Goodall's attribution of awe and spirituality to the Kakombe chimpanzees,[31] accusing the primatologist of anthropomorphising, he did so on the erroneous assumption that, as he put it, 'there is no meaning without language'.[32] Whereas in fact, '[m]eaning is more than words … [T]he fact that we are more aware of those times when we do think explicitly to ourselves in words … should not deceive us into believing that language is necessary for thought'[33] and meaning. The waterfall was meaningful enough to the chimpanzees to elicit a strong response,[34] while simultaneously, given how the brain works, it is likely that the chimpanzees engaged in some form of cognitive appraisal of the phenomenon. Like humans, other animals benefit from knowing and understanding the environment they live in, and their brains filter out information, sorting it into categories through experientially borne mindsets, allowing them efficiency in 'executing their lifestyle in their particular niche'.[35] Without the capacity to form concepts and categories, life for animals would be impossible because we would have to start from scratch with every subject, object, or phenomenon we encounter.

Once the existence of these processes in the animal brain is recognised, it is possible to appreciate that, like humans, other animals

---

31   Summarised in Goodall 2006.
32   Fisher 2005, 305.
33   McGilchrist 2009, 72, 107.
34   Moving rhythmically, throwing rocks and branches, and grasping hanging vines and swinging over the water.
35   Vallortigara et al. 2008, e42.

may encounter phenomena that resist automatic categorisation. When I say 'resist' I do not just mean that the phenomenon cannot be categorised because it is not known, not familiar, but also in the sense that it may be so broad that it does not fit, it cannot be forced into anything with limits, with boundaries, anything reductionist. This can then lead to the feeling of awe, which is a state of uncertainty that may call for a resolution. The resolution can take the form of appropriation by fitting it into an ideologically informed mental pattern, or it can take the form of a direct engagement with the encounter, a relational dance, in which the 'resolution' is not a closing but an opening to the process of engagement itself: a holistic embodied encounter that lies at the foundation of spiritual experience.

A fair amount of animals' (human inclusive) existence is thus characterised by various modes and intensities of cognitive closure, which is elemental for normative functioning. The invention/perception of deities, various cosmologies, social norms and similar conventions has functional value in this necessary yet reductionist process. Beyond – in fact, beneath – this level of consciousness lies the more holistic experiential reality, marked by intrinsic relationality: intra- and inter-organismically, animals' existence is inherently relational. The dynamic process of action and reaction produces tension that would ideally resolve into what could be described as psychic ('all inclusive') homeostasis – an ideal, balanced state of being that the organism strives for and that elicits the feelings of bliss often reported for spiritual experiences. Yet depending on the individuals' phylogenetic and ontogenetic legacy, such encounters between the experiential self and other agencies may also result in unfavourable feelings such as dread. Such a dance with animacy and its experiential potency clearly is not limited to the condition of being human, and nothing supports the assumption that dogmatic interpretations, which may result from cognitive closure, add anything to the intensity of the experience. One can interpret and contextualise the experience within a broader framework of one's life, but the experience itself is facilitated by the capacity to access the extra-linguistic, non-reductionist level that enables direct interaction, a synergic coupling with the nonself, which one can achieve once the mental templates are, to a large extent, silenced.

Awe is what I would refer to as an accidental source of spiritual engagement, in the sense that such engagement is not sought; instead it is triggered by an encounter, with a waterfall, for instance, or a landscape, that speaks to the animal in a particular way. It could also emerge from being present in a place that offers calmness and peace, and suddenly one may develop an overwhelming sense of integration within the self, a deep form of felt contentment as one becomes absorbed and absorbs the space with its tangible and intangible forces. It is these two forms of spiritual engagement that, in my view, mark the spiritual experience of nonhuman animals, and possibly many humans who, like the present author, do not engage in structured meditation or prayer. As opposed to accidental sources of spiritual engagement, meditation and prayer could be viewed as an intentional source as the individual intentionally seeks the connection, the experience.

Spiritual engagement, when the experience is positive, is not only an interesting subject for scholarly deliberations and a possible enhancer of religious belief, it also benefits animals' physical and mental health. I mentioned earlier that body-focused approaches, including meditation, are being increasingly employed in mental-health therapies for their overall benefits. In Chapter 2, I reported Coan's observation that bottom-up psychobiological regulation appears to be more efficient and less costly to an organism than top-down processes. Coan suggests that mindfulness meditation is characterised by the latter process. However, Chiesa, Serretti and Jakobsen have reviewed evidence and have come to the conclusion that in mindfulness meditation top-down regulation likely occurs in short-term practitioners, whereas in long-term practitioners the reverse appears to be true: regulation follows the bottom-up path.[36]

This observation is potentially significant for the consideration of animal (human inclusive) spirituality. As we have seen, psychobiological regulation plays a vital role in animals' lives, in infancy and beyond. Aside from regulation provided through social interaction, animals need to be able to self-regulate to various extents. If meditation, as a phenomenon that facilitates reconnection with the experiential self and extension of this self and consecutive merging

---

36   Chiesa, Serretti and Jakobsen 2013.

with the vitality of life and its intrinsic relationality, has such positive effects on human physical and mental wellbeing by influencing neuro-physiological structures and processes that humans in fact share with other animals, is it not safe to assume that other animals would benefit from it, too? Could that 'glimmer of infinity', to borrow Hugo's expression from this chapter's epigraph, reached when the experiential self engages in the dance with animacy, therefore be viewed not as a fad of the human 'divine' nature but as a biological imperative, on a par with alimentation, hydration and interpersonal relations? Nonhuman animals most likely do not train themselves in mindfulness meditation. However, is it not possible that they may have retained the capacity to engage in such meditative processes in a more spontaneous manner (a capacity that many humans, particularly those in industrialised societies, have lost and now have to re-learn), and perhaps to actively seek such experiences, or at least embrace them when they occur?

This resonates with Bataille's and Rilke's views from the beginning of this chapter. Clearly, as Willett points out, like humans, other animals also have to attend to more mundane things, such as searching for food, detecting danger, avoiding predators, and other cognitively and affectively demanding tasks that life brings along.[37] Bataille's *water* in the nonhuman animal world thus is not always shiny and blissful, with anthropogenic violence muddying it further. Nevertheless, the point I want to emphasise is that when the experience of connection, of bliss, does occur in nonhuman animals, and convergent evidence supports the hypothesis that it can, it is felt, experientially meaningful in itself, like hunger is, and satiety, thirst and hydration, social deprivation and union. It does not need a cognitively interpretative component – ideological or other – to give it meaning.

Spirituality and the self

To elucidate the disputability of the term 'self-transcendence' and to aid explorations of phenomena lying behind it, it is useful to consider the way the self is constructed and lived beyond the border of Western

---

37   Willett 2013.

dualism. The culturally informed diversity in self-perceptions within the human species combined with science-borne increased understanding of the implicit forces at work in the formation of animals' (human inclusive) experienced realities, may lead to the shattering of yet another wall that separatist human ideologies have erected to propagate human exclusivity – the wall that attempts to keep nonhuman animals outside the realm of the spiritual.

Self-relatedness

The seemingly obsessive preoccupation with the cognitive, reflective self in the West, arguably a product of left-hemispheric dominance among many shapers of Western culture, may have additionally been fuelled by the increasingly contested idea that self-relatedness is the output of a higher-order cognitive or meta-cognitive function. As discussed in more detail in the section entitled 'From neurons to neighbours' in Chapter 1, in this view organismic processes acquire experiential meaning only when filtered through tertiary, cognitive processes traditionally believed to culminate in (human) linguistic representations of self-awareness. This anthropocentric perspective surrounding self-referential processes has had significant repercussions for other animals who have largely been denied selfhood and the concomitant privileges granted to humans, who possess a declaratively conscious self. Moving the loci of self-relatedness to more ancient brain regions reinstates nonhuman animals into the realm of the 'fully conscious'. Further, Chapter 2, on intersubjective attachment and loss, demonstrates the potency of the implicit domain, which begins to shape through animals' interactions with the world before we are capable of autobiographic memories and before we have any control over what we allow to impact upon us affectively and in terms of thinking templates. The accumulation of implicit and explicit memories will tailor our 'integrative' self: the way we are and the way we see ourselves in the world and in relation to it.

In the domain of this more integrated self-relatedness such environmental–cultural shaping manifests as either an interdependent or independent self-construal. In the interdependent construal, typical for monistic traditions, humans are viewed as interconnected, and the focus of individual experience in this construal is 'the other' or the 'self-in-relation-to-the-other'.[38] In ecocentric traditions, such as those of many Indigenous people, this goes beyond the intra-human context and can exist in relation to land and other living entities resulting in an ecocentric (interdependent) self-construal.[39] However, in the typically Western independent construal, the self is viewed as

> a bounded, unique, more or less integrated motivational and cognitive universe, a dynamic center of awareness, emotion, judgment, and action organized into a distinctive whole and set contrastively both against other such wholes and against a social and natural background.[40]

In many Western cultures, in fact, individuals are encouraged to become independent, distinguish themselves from others, and discover and express their uniqueness.[41] In such typically left-brain grounded individualism the importance of the other and the social context is not negated, but the latter largely serve as criteria for reflected appraisal (how we think others see us) and verification and affirmation of the independent self.[42]

Western individualism stands out as an aberration given that the sense of self for humans and other animals emerges and continues to exist within a relational framework. Humans 'are immersed from the start, like other creatures, in an active, practical and perceptual engagement with constituents of the dwelt-in world', as Ingold phrased it, continuing:

> [t]his ontology of dwelling ... provides us with a better way of coming to grips with the nature of human existence than does the alternative, Western ontology whose point of departure is that of a mind detached from the world, and that has literally to formulate

---

38 Markus and Kitayama 1991, 225, 227.
39 e.g. Kirmayer, Fletcher and Watt 2008; Harvey 2005.
40 Geertz 1975, 48.
41 Markus and Kitayama 1991.
42 Markus and Kitayama 1991, 226.

it – to build an intentional world in consciousness – prior to any attempt at engagement.[43]

While interpersonal diversification is unavoidable since every individual (human or not) is a product of nature and the developmental micro- and macro-social environment, it is the bridging of differences – without denying the individual's needs – that ultimately enables a positive social dynamic upon which the wellbeing of animals depends. A minimised differential and a maximised sharing context facilitate the continuing process of information exchange inclusive of but not limited to the cognitive domain. This promotes symbiotic existence without dissipation or subjugation of the various participating selves. If individualisation and differentiation become expected as opposed to simply accepted and integrated, the individual risks becoming the projection of a highly abstract idealistic construct of the self that estranges the individual from oneself and from others, compromising the relational potentiality that is both intrinsic and vital to social animals.

The adopted self-construal also affects self-related processing on a neurological level. Zhu and colleagues scanned subjects from what they refer to as a typically interdependent (monolingual Chinese) and a typically independent (English-speaking Westerners) culture to test if self-relatedness extends to close others (in this case the mother).[44] They found that in the first case it tends to do so – the same brain areas are activated when processing the actual self and the mother – while in Westerners it does not. Based on that study and additional evidence (in congruence with the sense of self emerging relationally) Han and Northoff conclude that generally the self-/other-relatedness is best viewed as a 'self–nonself continuum rather than a self–nonself dichotomy',[45] with the degree to which the other is perceived as related to the self determining the cognitive and psychological distance on this continuum. There is currently no neurological data to compare human self-relatedness to nonhuman life in ecocentric versus anthropocentric

43   Ingold 2000, 42.
44   Zhu et al. 2007.
45   Han and Northoff 2009, 206.

traditions,[46] but differences could be expected, affecting relational potentials towards the nonhuman world and possibly also spiritual ones.

From self-transcendence to self-extension

In a public lecture entitled 'Becoming stillness' given in 2008 in Norwich Cathedral, visiting American Franciscan friar and Christian mystic Richard Rohr lamented the Western obsession with dualistic, absolutist thinking – an 'all or nothing' perspective – which he sees as having plagued Western Christianity, establishing or consolidating divides, and which appears to be absent in Asia.[47] Even the students he teaches in the United States who are of Asian descent, he noticed, seem to be more comfortable with paradox, less anxious to resolve it, compartmentalise it and filter it through the left brain, and more open to experiencing the world and their own personal creeds holistically, with their entire being. Rohr remains a Christian and retains a theological emphasis on the centrality of God and Jesus in his teachings; however, with his mystical practice and ideals he is an important voice in the deconstruction of the prevalent, divisive mode in Christianity that may not only exacerbate inter-group conflict but also affect people on a personal level, perhaps fuelling the sense of meaninglessness that some see as intrinsic to being human.[48]

Could this kind of meaninglessness source from a left-brain-induced distortion of the self in its relation to nonself and so-called self-transcendence? The independent self-construal engenders a level of detachment that may escalate to what I tentatively describe as ontological violence. The organism's emotional and other

---

46 Georg Northoff, email, 21 June 2016.
47 Here I am acknowledging the existence of cultural differences for the issue discussed without suggesting (and Rohr would probably agree) a romanticised view of a seamless union of body and mind in Asian thought and practice. While the mind is inherently embodied, as discussed in this book, a harmonious body–mind integration should be viewed as a result of growth rather than being automatic (e.g. Yuasa 1987), or as Samuel (2008, 11) puts it: 'The lotus of spiritual enlightenment [...] grows out of the mud of everyday life' for Asian and non-Asian people alike.
48 e.g. Fisher 2005.

interests are best served in a harmonious relational environment in which the self does not need to protect itself by way of detaching because the positive relational dynamic offers the necessary security for such protection to become redundant. The organism thrives in a relational context of positive mutual exchange, which benefits the organism on a bodily and more broadly psychological level, and, to borrow terminology from attachment theory, which becomes the organism's secure base for when disruption occurs, not only in childhood but throughout life. Disruptions are of course inevitable and result both from internal processes and engagement with external agencies, but the organism's propensity to establish and maintain favourable conditions appears in direct contrast with the alienation – exile – propagated by self-independency. Indeed, while avoidance of negative stimuli and conflict is desirable for organismic wellbeing, the independent self-construal may induce a rising of barriers between oneself and potentially favourable and complementary others, precluding, or at least substantially hindering, any kind of holistic engagement. When such an aberrational mode of being becomes a cultural norm it can lead to personal and social psychoses, which may be absent or at least attenuated in societies constituted of interdependent selves,[49] but which are lucidly illustrated by some pillars of Western philosophy. Søren Kierkegaard wrote clearly about it in the following journal entry:

> Deep within every human being there still lives the anxiety over the possibility of being alone in the world, forgotten by God, overlooked among the millions and millions in this enormous household. A person keeps this anxiety at a distance by looking at the many round about who are related to him as kin and friends, but the anxiety is still there.[50]

---

49 Juhl and Routledge (2014), for example, found that experimentally heightened death awareness increases death anxiety in people with a low (but not with a high) interdependent self-construal.
50 Kierkegaard 1980 [1841], 171.

The exile can be further exacerbated by cultural impositions propagating human superiority, and, by implication, inferiority and instrumentalisation of other animals and the rest of the world, and further even by ideologies that attribute transience to life on this planet, promising a more *real* life somewhere else. These factors could preclude full immersion in the immediate, embodied life on Earth, with consequent failure to appreciate the vibrant vitality of one's socio-natural environment and its countless relational possibilities, which may in some cases intensify feelings of loneliness and meaninglessness. Belief in God and/or gods and an afterlife does not preclude such immersion; however, the latter may be eased by the presence of pantheistic, panentheistic or animistic traits, which due to their greater permeability between the 'worldly' and the 'sacred' could aid in blurring hierarchical lines promoting division rather than inclusive relation. This process of immersion involves the abandonment of the logocentric self and a return to a more fundamental level of experience, in which the extra-linguistic self predicates relational dynamics as an embodied entity in synchrony with other bodies. This communion of bodies, which need not be human, enables what is known as 'self-transcendence', which is grounded and enacted on the above mentioned continuum of self–nonself.

Organisms, whether linguistic or not, have a basic capacity to distinguish between the actual self and nonself. Animals discern conspecifics as 'like self but not-self', which enables us to determine relative relatedness of surrounding organisms, predict behaviour (for example, friend or foe), and share resources. It also facilitates navigation more generally in the socio-natural environments.[51] The interdependent self-construal appears prevalent among the human population (West excluded) and is likely closer to other animals' perceptions of the self, particularly in its ecocentric variant.[52] On the journey to 'self-transcendence', the interdependent self-construal could be seen as a step ahead on this self–nonself continuum compared to the independent self-construal. Further 'merging' of the

---

51   Han and Northoff 2009, 206.
52   Bradshaw 2009.

self with other selves (other organic or nonorganic entities) promotes further 'transcendence'.

Such 'transcendence', as already indicated, is not so much about *transcending* as it is about *extending* (or *expanding*[53]) the self. In fact, when there is mention of self-transcendence the referent is the cognitive, reflective self, as one can never transcend the affective self and perceived embodied consciousness. The affective self is always there; it is the level of the self reached through meditation and other modes through which one speaks and is spoken to implicitly (hence more *directly*), such as music, communal chanting, dancing (hence the central part of these behaviours in religions), looking at stars or awe-inspiring landscapes, sitting in silence under the full moon chewing your cud (if you are a sheep), quietly with your close others. It is from this implicit communication with a *presence* (which invokes the body proper in its capacity for and inevitability of relationality with various stimuli) that spirituality emerges. Attending to the present moment aids in uncovering a multitude of relationality and relationability that may otherwise remain undetected. It is in this space of implicit relationality that self-extension materialises. Farb and colleagues explain:

> Narrative focus (NF) calls for cognitive elaboration of mental events, thereby reducing attention towards other temporally proximal sensory objects. In contrast, experiential focus (EF) calls for the inhibition of cognitive elaboration on any one mental event in favour of broadly attending to more temporally proximal sensory objects, canvassing thoughts, feelings and physical sensations without selecting any one sensory object.[54]

Momentary self-reference and the awareness of the psychological present, 'represented by evolutionary older neural regions' in the words of the authors themselves, 'may represent a return to the neural origins of identity, in which self-awareness in each moment arises from the integration of basic interoceptive and exteroceptive bodily sensory

---

53   Nelson 2011, 63.
54   Farb et al. 2007, 314.

processes'[55] in line with the previously discussed core self and implicit markers of self-referencing. The sense of presence that arises in this extra-linguistic, non-reflective inter- and intra-bodily communication – this affective dance with animacy – is at the core of spiritual experience. Albeit free of abstract, symbolic colouring, the dance itself is nevertheless marked by meaning-making; it *is* itself psychophysically meaningful. That is to say, it is felt, consciously experienced (albeit non-reflectively), and it informs associative, declarative renditions of experience.

The relationality fostered by the focus on the present moment has, for instance, direct positive effects on physical and mental wellbeing, with implications for illness vulnerability and mood and anxiety disorders.[56] Payne, Levine and Crane-Godreau outline the importance of attention to interoceptive and proprioceptive experience for trauma recovery. Their Somatic Experiencing® (SE) trauma therapy is centred on what they call the core response network (CRN). This network includes subcortical autonomic, limbic, motor and arousal systems, and is reminiscent of Panksepp's concept of the core self, Damasio's concept of the proto-self and Schore's concept of the implicit self, as Payne, Levine and Crane-Godreau themselves note.[57] The authors elaborate:

> A present fearful or stressful state is experienced in part as unpleasant interoceptive and proprioceptive feelings, including muscle tension, stomach tension, trembling, weakness, constriction, increased blood pressure (pounding pulse), decreased blood pressure (dizziness), increased or decreased heart rate, cold sweaty hands, hyperventilation, shallow breathing. Damasio terms these 'somatic markers', as they are bodily experiences of emotionally and viscerally activated states, consciously felt 'markers' of subcortical states. These somatic markers may activate memory traces that contain similar feelings. Such trauma-related memory traces may be partly or wholly inaccessible to ordinary conscious recollection, being procedural

---

55   Farb et al. 2007, 319.
56   Farb et al. 2007, 320, citing Segal et al. 2006 and Davidson 2004; see also Garland, Froelinger and Howard 2015.
57   Payne, Levine and Crane-Godreau 2015, 3.

or implicit rather than declarative and autobiographical. This means the person may not even be aware that old memories are being activated. Consciously recognized or not, the somatic markers connected to the old memories reinforce and augment the present fearful state in a runaway positive feedback loop, which can lead to terror, panic, rage, or shut-down.[58]

Conversely, positive implicit memories – memories devoid of explicit autobiographical awareness – can also be activated, leading to positive affective states, from exhilaration to more quiet but equally potent feelings of contentment, peace, immersion, communion, a sense of *home*, which are felt but not necessarily understood and contextualised. For example, the delicate touch of madeleines on the palate produced an overwhelming feeling of pleasure in Proust's *In Search of Lost Time*, invading the senses:

> something isolated, detached, with no suggestion of its origin. And at once the vicissitudes of life had become indifferent to me, its disasters innocuous, its brevity illusory – this new sensation having had on me the effect which love has of filling me with a precious essence; or rather this essence was not in me *it was me* [italics added].[59]

This moment of self-extension eventually becomes contextualised as Proust remembers his early encounters with the madeleines in his aunt Léonie's room on Sunday mornings. But, as indicated earlier, human and other animals' lives are replete with sensitised emotional systems and implicit memories imprinted through events and encounters, only some of which may be retrieved through conscious cognition, leaving others out of reach of the reflective self, yet alive and well from an experiential perspective, adding to the mystery matrix – the potential beauty and/or horror – of being alive.

Animal bodies move in this world with interoceptive and exteroceptive cues dictating our pace. The level of self-interdependence

---

58 Payne, Levine and Crane-Godreau 2015, 9.
59 Proust 2014 [1913], n.p.

and the ability to sustain an existential, relational openness may determine an organism's capacity to attend to the present moment and the given cues, allowing full immersion in life and relationality as opposed to condemning the self to a ghost town marked by paranoia and exclusion. Like human lives, the lives of nonhuman animals are both burdened and eased by implicit memories and organismic emotional and relational potentialities. Yet, arguably, they are less preoccupied with the left-brain rigidity that plagues the Western mind. This may enable nonhuman animals to be more focused on the present, suggesting that nonhuman animals may engage in the affective dance that marks spiritual experience more readily and regularly than many humans. Through such body-focused spiritual practice the world becomes alive and acquires agency, to borrow Harrod's phrase: *respondeo-ergo-sum*[60] (I respond therefore I am) to which I add *ergo es* (therefore you are) and *ergo est* (therefore he/she/it is). It is not only the subjects conventionally considered animate or sentient that possess agency; rather, by the capacity to elicit a response from me, the entire world possesses agency, from animal to plant to rock, manifesting, in Merleau-Ponty's words, an 'inexhaustible richness'[61] of perception and relation. This intimate moment of agencies touching each other gives rise to an implicit sacred whose experiential intensity does not depend on whether or not the (human) brain articulates it into an explicit sacred.

The affective dance at the foundation of spirituality can therefore be accessed directly by plunging into the relational dynamics offered by the living vibrancy of the world surrounding us and is mediated by implicit memories and other organismic functional properties. The narrative self can substantially inhibit such immersion. However, religion with its systemic conceptualisations may also aid the immersion by providing an orientational focus (for example, intimate communication with God) for the reflective self to 'lose' itself and give space to the experiential self to conduct its affective dance with the *presence*, which has by this stage become 'embodied' in the sense that it has lost its abstract essence and is able to interact directly with and within the organism on an experientially conscious albeit

---

60   Harrod 2016.
61   Merleau-Ponty 1962 [1945], 279.

non-reflective basis. NDE and other potential markers of spiritual experience are, at least in the human brain, vulnerable to interpretations of grand proportions. There is no doubt that the associative reality at the base of cultures and organised religions gives rise to symbolic meaning-making and meaningfulness[62] which penetrate deep into the individuals' being and inform the explicitly symbolic sacred with notable affective implications. For instance, the emotional connection with God may lead people to develop secure or insecure attachment relations with it (for example, loving or distant), mirroring those with caregivers and other conspecifics.[63] Nevertheless, such explicit symbolism is not critical for meaning-making nor is it imperative for spiritual engagement; both can, and do, arise directly through the organism and its implicit, non-reflective self, with the value differential between direct (non-reflective) and indirect (reflective) access to the world 'beyond' likely stemming largely from anthropocentric prejudice.

Cognitive closure is exacerbated in human organised belief systems, but as noted earlier, all animals engage in this process to various extents in their attempt to understand and attain a sense of control over or manageability of the various phenomena in their environment. As Guthrie, who has written extensively and convincingly on the topic of agency detection and ego-morphism in the animal world in relation to religion,[64] points out, and as has been argued in this book in relation to spirituality:

In the search for an explanation of religion, I believe, we have been beguiled by symbolism and misled by a false sense of human uniqueness. As a result, we have forgotten a vital need that we share with other animals: to interpret an ambiguous world and to discover real agents hiding in it.[65]

---

62 e.g. Fisher 2005.
63 See, for example, Granqvist 2002; Reinert and Edwards 2009; Granqvist et al. 2012; Counted 2016.
64 See, for example, Guthrie 1980 and Guthrie 1993.
65 Guthrie 2002, 61.

So-called habituation is one such process whereby increased exposure to a phenomenon that is benign but may initially be perceived as dangerous due to its strangeness and foreignness, and/or due to its symbolic properties, gradually undergoes accustomisation (a process that requires the affected animal's re-evaluation), losing its fear-inducing potency. *Better safe than sorry* – the best strategy when in doubt[66] – does not derive from a human-exclusive brain manoeuvre (except perhaps for its syntactic rendition), it is a basic survival philosophy that we share with other animals. Attribution of agency and intentionality to seemingly unlikely candidates, as naïve or 'primitive' as it may sound, makes perfect sense outside the sterile world of absolute control of the WEIRD population. Outside heated and air-conditioned apartments, in the absence of Biokills and snake removalists, in places where water does not flow out of taps and food is not readily available on supermarket shelves, a different kind of reverence may arise from the immediacy of the surrounding agency, which can perhaps only remotely be grasped by anyone partaking in this reality as a guest but whose life is not immediately dependent upon it. The wind may usefully disperse plant seeds, promoting abundant food resources, but the wind may also diffuse sounds and smells, disorienting the listener and increasing vulnerability to predators. Water may soothe the thirsty but may also drown them. As such, wind and water, and other phenomena, are real and complex agents impacting upon animals on an affective level and certainly well worthy of their reflective consideration. Whether such considerations extend to applying theistic or similar properties to or behind these phenomena is of secondary (if any) relevance; the dance with the wind and other 'spirits' animals engage in is undoubtedly experientially meaningful but also conceptually challenging for them given that such relationality is marked by a(n) (in)tangibility unlike anything organisms experience when dealing with an immediate threat of a foe or the soothing care of a friend. In the latter case, both friend and foe are to a great extent graspable in their intentions and actions, aiding the choice of an adequate response.

---

66   Guthrie 2002; see also Guthrie 1993.

Furthermore, animals' associative powers – symbolism, often considered to be a human-only domain – play an important role in coping with the ever-changing and intrinsically agentic nature of the world. For instance, in various animal species stripes appear to raise the red flag. I first became conscious of this fact in an attempt to put striped jackets on freshly shorn sheep to protect them from the cold mountain air of early spring. Striped jackets were a result of the limited selection of patterns in the pet-store rather than a premeditated decision. While the sheep like to exercise self-determination and largely refuse to be interfered with except for friendly interactions such as gentle stroking, their response to the jackets exceeded any expectation. Initially I suspected they may have been playing games with me, as they sometimes do. It was only after I had managed to install the jacket on one of them, which set him running madly around the paddock, desperate to get away from the 'monster' on his back while another one, seeing his reaction, started to climb up the wire fence determined to escape, that it became clear to me that they were playing no games but their response reflected genuine and absolute terror. Arnold Chamove observed the response to stripes in shelter dogs, and found increased levels of fearfulness and decreased friendliness in dogs exposed to a human wearing striped clothing compared to clothing without a pattern.[67] In the case of dogs, the response could be understood within the context of aposematism,[68] which describes colour (but also sound and smell) adaptations of certain prey species to deter potential predators by signalling the prey's unpalatability, toxicity and/or other adverse effects should the predator be considering attack. For sheep, as herbivorous prey species, however, stripes and other striking patterns and colours may symbolise predator attack rather than prey defence, much like the smell of a cat would put a rat in a state of alert. Such symbolic abilities, once again, hint to the richness and complexity of nonhuman animals' minds and being, so often ignored or downplayed in their significance.

Taken together, the capacity – indeed the imperative – of direct, experientially conscious communication with phenomena in the

---

67 Chamove 1997.
68 The term was coined by Poulton (1890); see also https://bit.ly/3jnInYw.

environment determines the affective dance with animacy as a felt interactive presence on a self–nonself continuum, which I see as the foundation of spiritual experience for animals (including humans). The rich and complex sensory realm of the living world interacts with animals' phylogenetic and (implicit and explicit) ontogenetic memories, eliciting positive or negative affective states and related behavioural responses, such as in the previously mentioned case of the chimpanzees.

When a phenomenon is perceived as adverse, the experiential self may feel the need to break the continuum, isolate itself on this continuum, to preserve psychic homeostasis in the face of a potentially disintegrating force. It may be that in such adversity there is a greater urgency for a reflective grasping and 'ownership' of the experience, the cognitive closure that may determine religious imagination. Phenomena eliciting positive feelings, on the other hand, may expedite merging of the self, inducing self-extension. Ultimately, this is beneficial for the body-mind as it promotes homeostasis, and is reminiscent in several ways of the positive exchanges within the attachment dyad. An individual may be inclined to expand this experience through continuing immersion rather than confining it through reflective closure, even though the latter likely, to various extents, represents a normative categorisation process of the brain following any kind of experience for human and nonhuman animals alike.

Place attachment and the roots of spiritual relating

> Before the gods existed, the woods were sacred, and the gods came to dwell in these sacred woods. All they did was to add human, all too human, characteristics to the great law of forest revery.
> — Gaston Bachelard, The Poetics of Space[69]

---

69   Bachelard 1994 [1958], 186.

'Inhabited space transcends geometrical space,' writes Bachelard in his exquisite exploration of dwellings and the broader spaces accommodating them.[70] We live and breathe spaces, we create them and they create us. Space is so integral to being that it becomes elusive. Although all animals likely exhibit place preferences (and aversions) based on phylogenetic and ontogenetic factors, research on place attachment continues to struggle with conceptual and methodological problems deriving from the intangibility of a phenomenon, which is nevertheless, and not unlike many other ungraspable phenomena, composed of tangible entities with geometrically and otherwise defined and describable features. A more general definition of place attachment, which would bring higher homogeneity to this field of inquiry and further the theoretical development of the concept, has yet to be agreed upon.[71] With place attachment being a continuation of community studies, there is a strong tradition of focusing largely and principally on social aspects of place attachment – the extent to which a place facilitates inter-human sociality and community identity.[72] This approach has been contested by some – most notably perhaps by Richard Stedman[73] – who are disinclined to view the physical space only as a container for human social interactions. Instead, as suggested throughout this chapter, from the organism's capacity – indeed, the organism's imperative – to communicate with the environment and entities within it emerges a relationality that creates experiential (albeit not necessarily always reflectively elaborated) meaning, whereby meaning, as summarised by Johnson 'is not just a matter of concepts and propositions, but also reaches down into the images, sensorimotor schemas, feelings, qualities, and emotions that constitute our meaningful encounter with our world'.[74] Paul Morgan draws parallels between interpersonal attachment theory and person–place attachment, proposing a developmental model of place attachment that, like interpersonal attachment, is a result of the exchange between

---

70 Bachelard 1994 [1958], 47.
71 Scannell and Gifford 2010, 2.
72 Summarised in Lewicka 2011.
73 Stedman 2003.
74 Johnson 2007, xi, cited in Lewicka 2011, 221.

the child and the physical environment as an interacting presence. Rather than understanding the motivation to interact with the environment as sourcing entirely from within the child, Morgan, citing Striniste and Moore, views such motivation as 'both a quality inherent to the child, which determines how the child will use the environment, and a quality of the environment, which has the potential to draw the child's involvement'.[75]

Through repetition this relational dynamics of arousal-interaction-pleasure generates an affective bond between the animal and the place, consolidating into an unconscious internal working model informing future relationships with place. Place thus has the capacity to soothe and, in Kalevi Mikael Korpela's view, much like intersubjective attachment dyads, place may function as an external regulator mediating psychic balance.[76] If such consolidation does not occur in childhood, place attachment later in life may be weak. However, there is currently not enough data to support this claim. In fact, given the inevitability of interaction with places and given that extra-ontogenetic factors may contribute to animals' affective responses to them, there is reason to believe that, like in the area of intrapersonal attachment, the organism remains open to place-attachment potentialities and affective reorganisation all through life with attachment emerging if the relational transaction secures the kind of multimodal support the organism needs and/or expects.

Fashion issues aside, the popularity of home-décor media and the financial prosperity of this market with an estimated gross of US$65.2 billion per year in the U.S. alone,[77] suggest a strong human inclination to concentrate substantial effort and other resources (when they can afford it) on making their residential hub not only mechanically liveable but also relationally competent. Relational competence in this context would imply the place's capacity of implicit communication with the dweller, of arousal and soothing, of promoting or supplanting moods, creativity, concentration, relaxation – features that determine the difference between a house and a home. In the choosing and shaping of

---

75 Morgan 2010, 14, citing Striniste and Moore 1989, 25.
76 Korpela 1989.
77 Summarised in Graham, Gosling and Travis 2015.

residential spaces, people may be driven not only by explicit tastes and preferences but also by unconscious factors reflecting emotional ties to certain features, sourced from past experience with them, as Graham, Gosling and Travis explain.[78] For instance, people may unconsciously incorporate features from a beloved grandmother's house – those madeleines of interior design. However, despite the obvious psychological significance of residential spaces, the authors lament the lack of empirical research in the area even though it has received substantial attention by theorists and practitioners, including Jung,[79] and Bachelard. The home with its sense of familiarity, cosiness and perceived protection indeed has the capacity to function as a bioregulator as Korpela suggests. On a cognitive-affective level, much as it can be the case in interpersonal relations, the home can become part of the extended self, and if the home is violated, for example by burglary, people report qualitatively similar albeit less intense psychological distress compared to victims of direct physical violations.[80] However, human-made buildings and their adjacent yards are only part of the composition we may call home. The latter encompasses a much broader space, the limits of which are not always easy to identify. Research into environmental degradation in and surrounding human habitats (for example, due to mining operations, land clearing, and similar) also records psychological distress,[81] which Glenn Albrecht named 'solastalgia' (as mentioned in Chapter 1) and described as:

> the pain experienced when there is recognition that the place where one resides and that one loves is under immediate assault (physical desolation). It is manifest in an attack on one's sense of place, in the erosion of the sense of belonging (identity) to a particular place and a feeling of distress (psychological desolation) about its transformation. It is an intense desire for the place where one is a resident to be maintained in a state

---

78   Graham, Gosling and Travis 2015.
79   Reported in Graham, Gosling and Travis 2015.
80   Droseltis and Vignoles 2010.
81   e.g. Higginbotham et al. 2006.

that continues to give comfort or solace. Solastalgia is not about looking back to some golden past, nor is it about seeking another place as 'home'. It is the 'lived experience' of the loss of the present as manifest in a feeling of dislocation; of being undermined by forces that destroy the potential for solace to be derived from the present. In short, solastalgia is a form of homesickness one gets when one is still at 'home'.[82]

Rogan, O'Connor and Horwitz, on the other hand, report compelling testimonies of farmers who had themselves (along with their ancestors) been part of such an assault.[83] With the 'golden past' in mind, one that they had never themselves experienced, at least not in the place in question, the degradation was nonetheless obvious, and upon realising the consequences of the long-held belief in the 'normality' of the utilitarian approach that led to such degradation, they changed their ways. A turning point for a farming couple interviewed, for example, was their experience with wind erosion, 'one of the worst experiences we had', they are reported to have said.[84] Another farmer admits that he had always seen the environment as a means for agricultural gain: 'I didn't see the clearing really, because that was good, because we were producing something where before nothing was produced ... and this is what farmers do, they clear the land', until one day flying back home, 'it suddenly struck me that what I had in my hands was the spoils and I was one of the spoilers that was making it look like that out of the window. It was just so graphic that it was mind boggling.'[85]

Unguarded instrumentalisation of nature following the emergence of agriculture, but particularly over the past few centuries with technological advancement and unprecedented human population growth, has contributed substantially to the degradation of the planet. This is now backfiring, not only in terms of geophysical changes with their adverse effects (such as erosion mentioned above), pollution of vital resources such as water, extreme temperatures, etc., all of which

---

82 Albrecht 2005, 45.
83 Rogan, O'Connor and Horwitz 2005.
84 Rogan, O'Connor and Horwitz 2005, 151.
85 Rogan, O'Connor and Horwitz 2005, 151.

may affect human (and of course other animals') physical health as well as community wellbeing.[86] This anthropocentric, self-appointed role of dominion (or 'guardianship' as we prefer to call it these days) over the rest of nature has uprooted humans from their evolutionary cradle to the extent that it appears to be quite directly impacting on their psychological health. The field of ecopsychology in particular has been concerned with the troublesome relations between humans and the rest of nature. Congruent with the biophilia hypothesis[87] whereby humans tend to feel an urge to affiliate with other life forms, this field explores the interdependency of humans and the rest of nature and the benefits a greater appreciation of this synergy and reciprocity by humans would bring to both.

Like other animals, humans have also evolved in a naturalistic setting with large-scale urban spaces being a relative novelty in our species' history. Regardless of human ontogenetic experience and reflective conclusions, human bodies and psyches recognise this intrinsic connection and may ache if deprived of it. City dwellers, for example, may think they are habituated to stress-inducing urban environments, but their bodies and brains may continue to react to the stress without the human being reflectively conscious of it, as Bratman, Hamilton and Daily observe.[88] An indication that this may be the case can be found in the numerous reports of therapeutic properties attributed to exposure to 'nature', including a sense of connectedness, belonging, and an extended (eco) self, which tend to be labelled as spiritual.[89] Even the smallest exposure appears substantially advantageous: Ulrich found that rooms overlooking hospital gardens positively affect patients' recovery rates while Moore notes that prisoners benefitting of similar views exhibit lower rates of sick-calls.[90] Koga and Iwasaki, with a focus on the tactile consciousness and simultaneous repression of the visual dimension, examined psychological and physiological effects of touching plant foliage

---

86 Berry, Bowen and Kjellstrom 2010.
87 Wilson 1984.
88 Bratman, Hamilton and Daily 2012, 123.
89 e.g. Snell and Simmonds 2012.
90 Ulrich 1984; Moore 1981 (both reported in Hinds and Sparks 2008, 110).

compared to other textures/materials. The results, based on the experimental human subjects' reported impressions and on observations of their cerebral blood flow, show that touching plants has a reflectively unconscious (but of course experientially conscious) calming effect on people, leading the authors to conclude that 'plants are an indispensable element of the human environment'.[91]

While a substantial part of the literature addressing the human and more-than-human nature relationship appears to promote humans' respect for the rest of nature, the instrumentalisation of nature implicitly and most likely unconsciously propagated in these writings is disturbing. The notion of nature as a place to go to in order to de-stress, to get away from the daily capitalist, consumerist life with its overwhelming contribution to the destruction of nature, and recharge one's internal batteries for better coping with the stress-inducing 'ordinary' life, gives little consideration to the fact that any such visit to 'nature' is in effect an invasion. It is an intrusion into spaces of other animals (including us, the local human residents), into *their ordinary* life, and as such causes additional stress to *them*. Nature is not an empty cathedral awaiting humans to come and unload their burdens, even though at times it may appear so given that animals tend not to put themselves on display for the 'wilderness'-hungry human visitor. Instead, nature is a fully inhabited space, home to communities and individuals who have been pushed to the limits (physically and psychologically) by the destructive powers of the ever-expanding human population, urbanisation, industrialisation, and more recently eco-tourism and eco-therapy. Walking into other animals' homes and backyards will cause a certain amount of anxiety to the dwellers as any novel situation inevitably does. It will disrupt the animals' current activity (whatever that may be at the time of the intrusion: acquiring food, constructing a home, educating their children, solving a personal dispute, resting after a hard day's work, etc.), forcing animals to attend to the intrusion because failure to do so may cost them, their families and their friends their lives. A recent disturbing report shows that anthropogenic disturbances, which include anything from hunting to hiking, are forcing free-living nonhuman animals around the world to become more nocturnal. Since it has

---

91   Koga and Iwasaki 2013, 1.

become increasingly harder to avoid humans in space, they have had to adjust their lives to avoid us in time.[92]

The 'wildness' cherished in animals is far from the unstructured randomness that it is often credited with. Statements such as Cookson's: 'Wildness in humans is usually seen as disruptive, while wildness in animals is essential to the health of their ecosystems' and 'If humans do not make the grade needed to use wildness constructively, then wildness becomes an otherness that is shunned',[93] disregard the ecological and social sophistication, which is, as emphasised previously, characteristic of other animals' communities. Like human communities, other animals' communities are also regulated by social norms and codes of conduct with aberrations exceeding acceptable levels likely to be penalised in one way or another. And just like intra-community social norms provide a certain grade of stability and security in animals' lives, so does familiarity with the residential space and the conspecifics and others in it, with strangers and the unpredictability accompanying them causing due concern. 'I see, now, "the wilderness experience" almost as an escape from the cultural reformation work needed, from reforms needed for survival of life on the planet,' Robert Greenway admitted in an interview.[94] He continues, 'Wilderness-as-therapy seems (in my more cynical or frightened moments) as an indulgence – an experience primarily available to a relatively wealthy tiny minority of the planet's human population.' In the absence of such a reform, humans and other animals tend to their primordial needs for psychic wellbeing as best as their circumstances permit it, travelling through space, touching wildness and being touched by it, often without much reflective awareness of it.

When psychologist Peter Kahn goes on his writing retreat, leaving human society behind for two months to write his academic book, it is not the wildness of his unshaved face and his unbathed body that makes his experience so special, even though he seems to have enjoyed that, too.[95] Rather, it is the wildness that emerges following the

---

92 Gaynor 2018; Gaynor et al. 2018.
93 Cookson 2011, 188.
94 Greenway 2009, 48.
95 Kahn 2009.

blurring of time frames and schedules that many humans are slaves to in their daily lives that provides him with the opportunity to be present and with the time 'to notice', as he put it, and the silence, which aided both. Kahn writes:

> Odd how the silence I described on my first night here was pounding my head. It was like I was coming off drugs and noise was my drug. This morning there is lovely sound in this light rain, and the blessing of the silence between the drops.[96]

When silence is allowed into a space, the organism opens up to modes of communication and understanding – of relating – that may be suppressed or obfuscated in an environment with noise as a prevalent theme. In our evolutionary past and in the rest of nature, sound tends to be used sparingly and meaningfully; attention to detail draws very fine lines between life and death: is the rustling foliage just signalling a cool change or is it aiding a predator's disguise? The world is populated with 'spirits' (agencies), and other animals are as aware of them as humans are, regardless of whether they develop stories around them or not.

I was able to observe my own change of perception and increased attention to detail over the four months I spent in the paddock and the adjacent woods most of the daytime when I was raising Orpheus-Pumpkin, the rescued orphaned lamb mentioned earlier. This area is home to some of the world's deadliest nonhuman animal species, such as the eastern brown snake and the funnel-web spider, both of which live on and around the property; therefore, while sitting on the ground, barefoot most of the time, caution was in place. As Guthrie argued in his variation of Pascal's wager in relation to anthropomorphising: 'it is usually better to err many times by applying them [human-like models] when they do not obtain than to err once by failing to apply them when they do'[97] – it is better to be cautious and take perceived agency and/or intentionality seriously since failure to do so may one day come at a great cost.

---

96   Kahn 2009, 42.
97   Guthrie 1980, 190.

While waiting for snakes and spiders, one becomes sensitised to all life around, as minuscule as it may be physically, particularly anything out of the ordinary, the unknown and hence unpredictable. Following this period of close interaction with the ground and bushes on the property, I took a trip to the city; walking down the pavement I saw an unusual creature move in the middle of it. I stopped, convinced at that stage that it was an animal of some kind, but upon closer inspection I realised it was a leaf that had been moved by a gust of wind. I smiled, only half pleased.

Just like one learns the rules of urban traffic, one learns to navigate other shared spaces. In doing so animals discover spaces that are best able to communicate with them and fulfil their psychobiological needs, which may result in place attachment. Conversely, communication with local agencies can be less positive, resulting in aversion.

The wild rabbit who 'domesticated' himself and settled in our garden would spend most of his time watching the rest of the world from under the lemon tree. The sheep spend the day grazing around the property but at night, or even during the day when they feel like resting, they always come to the same spot on a small plateau within the paddock. Elevated areas with a good view over the rest of the habitat and a higher chance of early predator detection are a rather common choice as a resting place in this species. Within this plateau they exhibit even more specific place preferences and it is not uncommon to see a sheep whose spot has been occupied by another sheep nudge (or butt) the intruder to encourage them to move. When it rains the sheep take shelter in the barn, except for Jason. Rain speaks to Jason differently than it does to the others. Jason stays outside, lying under the rain with his face up and eyes closed. When the rain is heavier he moves under the tree; when the rain gets even heavier he joins the others in the barn.

Place attachment in humans is believed to not develop until the age of five at the earliest, but more probably later and 'certainly' after human-to-human attachment.[98] In my view, there are several problems with this assumption. The first and most obvious one is that relying on human self-reports can only reflect the contents of explicit memory and not that of the implicit memory storeroom, which has an equal or

---

98   Morgan 2010, citing Sobel 1990.

greater impact on animals' perceptions, feelings, choices and decisions in life compared to the explicit contents. If we cannot recall place attachment from an earlier age it does not mean that place attachment was not present, or that it did not, at least, begin to form far earlier than our recollections suggest, as Morgan also observes,[99] probably earlier than, or perhaps alongside, intersubjective attachment.

In fact, unlike lambs and other precocial infants who pop out of the womb (or egg – giving birth to live offspring is basically a delayed form of egg-laying[100]) and start following the caregiver, many animals (humans included) do not consolidate intersubjective attachments until they become capable of moving and getting lost. Nelson and Panksepp suggest that intersubjective attachment may have evolved from ancient mechanisms regulating place attachment, energy balance, thermoregulation and pain perception, with the first three contributing primarily to evolving mechanisms for the appreciation of social presence, while mechanisms for pain perception play a greater role in the development of mechanisms for distress related to social absence.[101] Panksepp never pursued research into place attachment as the origin of intersubjective attachment nor was he aware of anyone else that did so.[102] Nevertheless, from a neurochemical perspective, the endogenous opioid system with its capacity to modulate physical and social pain and pleasure plays a central role in mediating attachment relations, with the latter more likely to develop when a beneficial balance felt on the level of the experiential self is reached. In place-preference conditioning, for instance, place preference is achieved by pairing a place with opiate administration.[103] This does not tell us much about non-conditioned processes of place attachment, but it suggests that a place with the capacity to trigger an animal's opioid system would be a suitable candidate for an attachment relationship, with mothers being one such place.

---

99 Morgan 2010, 20.
100 Goodson, Kelly and Kingsbury 2012.
101 Nelson and Panksepp 1998, 438; see also Panksepp 1998, 265; MacDonald and Leary 2005.
102 Jaak Panksepp, email 18 June 2016.
103 Nelson and Panksepp 1998, 438; see also Prus, James and Rosecrans 2009.

The line between subject and place appears far more blurred than is generally recognised. Whether subject is place or place is subject may be difficult to establish, or rather the two 'states' may be better viewed as fluid, taking the form of one or the other depending on the circumstances and viewpoint in any particular moment. Before having the opportunity and the capacity to interact with other animals, an animal lives and develops in, and interacts with, space. The womb, or the egg, is not a stimuli-free void, a sensory desert animals have to go through before 'real' life starts. Real life starts at conception, and everything that happens afterwards affects the organism, and it continues to do so until death. Both the womb and the egg represent a vibrant environment with sensory input the foetus is influenced by and responds to, including endogenous (within the mother's body) and externally generated sounds.[104] As mentioned in Chapter 1, in human foetuses, for instance, these sounds influence the post-natal organisms' phonetic perception. Human babies react differently to familiar (native) and non-familiar (foreign) vowels,[105] while domestic chicks communicate with both siblings and the mother while still in the egg. Before having the opportunity to pair a voice with a face, the animal interacts with cognitively intangible entities, a relationality that appears to be the default mode of the experiential self.

The experiential self is in constant communication with the 'spirits' in the environment, which are responding to it and elicit responses from it. There is a sensed presence before these agencies are associated with a particular face, smell and other specificities. Even when they do become associated with an embodied entity the mechanisms underlying the relation remain implicit, 'spirit'-like, sensed but largely escaping capacities for reflective articulation because such interaction does not address one level only, but a multitude of levels reflecting various organismic needs. Once out of the womb (or egg) the organism continues its rich communication with space. Attachment figures (both in infancy and later in life) are primarily places (albeit very specific ones) with the capacity to offer the multimodal support the organism needs. This is why the principal

---

104 Griffiths et al. 1994.
105 Moon, Lagercrantz and Kuhl 2013.

tragedy of losing a significant other lies not in the philosophical realms of abstract questions concerning one's own mortality or immortality that may emerge following the loss, loss is painful primarily because the sudden absence results in a sudden withdrawal of this multimodal support the survivor enjoyed in the complex space of the embodied individual lost. On this fundamental experiential level, grief can be equally acute for human and nonhuman animals.

The individual as space is perhaps most obvious in the infant–mother dyad in altricial species. Before mutual intersubjective attachment develops – before the infant recognises and bonds with the primary caregiver – essentially the infant is interacting with a space. Unlike some other forms of space (such as a room or a garden), this space (the mother/caregiver) is a very active one and it intentionally seeks interactions with the infant, providing stimulation as well as regulation of the infant's internal states, as discussed earlier. Despite the intentional interaction and care provided by the caregiver, communication within the dyad occurs on an implicit level. As such, factors outside the caregiver's will and intentional focus affect the interaction and determine the success of the regulation or lack thereof. These factors include the caregiver's own attachment style, stress reactivity, dispositional anxiety and other components of the micro-cosmos embodied as a primary caregiver that communicates with the multimodality of the micro-cosmos of the infantile organism. Adult attachment relations reflect a similar multimodal exchange engaging a conglomerate of properties and activity, which is better understood as a more or less integrated space rather than a singular entity, if any such thing exists in the first place.

If an individual can in essence be viewed as a space, can space/place be perceived as individual, a personified (though also multimodal) entity?

As part of the therapeutic process, Siegel encourages his patients to imagine a place from the past or present where they felt safe and happy.[106] Studies in separation distress reveal that a familiar environment may reduce separation distress vocalisations in animals.[107] Further, oxytocin, the mammalian neuropeptide with

---

106 Siegel 2011 [2010].
107 Panksepp 1998, 265.

homologies in other taxa, which plays an important role in social interactions and increases the sensitivity of the opioid system, when externally administered to animals may increase social bonding, but only if animals have not had the chance to fully habituate to the test chambers. If such habituation occurs, oxytocin is ineffective, which may mean that the existence of a reasonably strong place attachment may hinder the formation of new intersubjective bonds.[108] This suggests that the line between subject and space may indeed be far more blurred than generally thought. In a similar vein, while oxytocin may promote prosocial behaviour towards in-group members, it has the opposite effect towards out-groups.[109] In research, nonhuman animals' place attachment tends to be conceptualised as so-called territoriality. Like in the exploration of human place attachment where place has largely been viewed as a container for intersubjective and intergroup social interactions, territory (and territoriality) is also usually explored through the lens of inter- and intraspecific subjective functionality (for example, space providing mating opportunities, etc.), rather than focusing on the direct relation between individuals and their space.[110] Such relation may indeed be elusive, at least compared to the much more (observably) vibrant animal-to-animal interactions, with (other) places likely appearing largely static to an external observer, but the organism is nevertheless in constant interaction with them and this interaction, this relation, informs the organism's state albeit in a largely reflectively unconscious way.

Just like a singular animal (human or other) has the capacity to speak to the sensory realm of another animal leading to either attraction or aversion, other spaces are also in communication with individuals through touch, sight, smell and other sensory modalities. Phylogenetic knowledge may broadly define needs and preferences; it may explain, for example, why wilderness has a powerful calming and

---

108 Panksepp 1998, 252.
109 Panksepp and Biven 2012, 39.
110 The discourse concerning cage/pen enrichments for captive animals is a clear exception since it aims at providing a more naturalistic setting in recognition of the multimodal support such setting may offer to an organism (e.g. Balcombe 2006).

so-called spiritual effect on people, why touching plants is beneficial as the Japanese study showed, and why sheep choose the highest spot in the paddock to relax and watch the stars on a clear night. Ontogeny defines such preferences further.

* * *

Of the many attributes that have over time been used in attempts to distinguish the human animal from other species, spirituality is one, jealously guarded by those whose humanity may feel under threat if it turns out that the sacrificial lamb has equal, or greater, spiritual depth compared to her killer. The presumption, even among those who identify as 'secular', that spirituality is both unique to humans and also somehow 'beyond' the human – a kind of 'divine' feature of the human, linking the human to the 'divine' – has led to the view that spirituality is something above and beyond the cognitive. Beyond this, however, many Western commentators do not appear to know where to go. Because of the meta-physicality of such presumption, the last place most[111] have looked into to further our understanding of spirituality is the body and the more ancient brain regions that humans share with other animals. There are ethical implications in maintaining this ideological divide. Concerned humans may choose in this manner to continue to separate themselves conceptually from the rest of the world but for as long as humans exist we will impact upon that world.

Spiritual experiences may have significant psychobiological value for animals, both human and other. Anthropogenic violence, such as incarceration with its inherent deprivation of species-specific natural sensory and other input, or the destruction of the planet, which is as much their home as it is ours, may prevent or obstruct nonhuman animals' engagement in spiritual experiences. This may substantially impact upon their wellbeing. On top of the grief they experience at the loss of significant others, many animals may also be grieving the

---

111 However, the tide is turning also in religious studies where the central role of the body and sensory perceptions in life, including religious lives, is increasingly being recognised. See, for example, Grieser and Johnston 2017; Promey 2014.

loss of an ontological normative that has been tailored through a long evolutionary path and that humans have more recently come to manipulate for their own purposes. Further, such violence with its rampant objectification and instrumentalisation hurts not only the nonhuman animal direct victims but also those humans who have *seen* and cannot *un-see* and look the other way.

5
Grief at a distance: Humans grieving unknown animals

I hear their screams and witness their fear and suffering in hundreds of places including slaughterhouses, industrialised farms, darkened sheds, open paddocks, feedlots and inside transport trucks and ships on four continents.
— Patty Mark, *Humane Myth*[1]

Hope is not something you feel. Hope is something you do, regardless of how you feel.
— pattrice jones, 2017[2]

A phenomenon that has gathered momentum in recent years within the animal rights and liberation movement is public vigils for nonhuman animals, particularly for the victims of organised anthropogenic violence, such as animal agriculture, victims who are mostly personally unknown to the griever. These vigils and related public demonstrations in honour of the innocent victims may be perceived as yet another innovative method for bringing the plight of nonhuman animals to

---

1   Mark n.d.
2   Jones 2017, n.p.

the attention of the broader public, but in fact the implications are far more profound. On the one hand the vigils challenge the precarious delimitations of grievability, reflected in the mindset and practices of the larger part of anthropocentric and consumerist societies. On the other hand they provide social and emotional support to the growing number of individuals who live with such grief but since they are rarely understood they remain marginalised.

I already noted that a substantial amount of advocacy, both regulatory and abolitionist, revolves around painism.[3] This is justifiable given the amount of physical suffering nonhuman animals endure at the hands of humans. The suffering often escapes the attention of the consumer, in part because agribusiness and related industries are legally protected from public scrutiny. The footage of sheep in overcrowded pens on live-export ships, half-buried in their own excrements and cooking from the inside out that shocked the Australian nation in April 2018 was obtained illegally as these operations prohibit filming on board.[4] It is equally rare for conditions on farms and in slaughterhouses to be broadcast publicly. However, with footage aplenty (albeit in most cases also illegally obtained), the reason for such neglect must lie elsewhere.

On farms, too, animals may be dying in conditions similar to the ones on the so-called death ships. The hens from a free-range egg facility in Lakesland, New South Wales, Australia, are a case in point. In June 2018 thousands of hens were discovered, starved and water-deprived, struggling to stay alive among the cadavers of comrades who had succumbed to what turned out to have been an illegal act of purposeful deprivation known as 'forced moult'. Forced moult describes the practice of withholding food and water to force hens to moult, leading to increased productivity, at least in those who survive.[5] Following inaction by both the RSPCA and the police, concerned members of the public stepped in to rescue the hens, an operation led by NSW Hen Rescue, a reputable charity that rehabilitates and rehomes former laying hens, in conjunction with activist group Legion DX Sydney. The outcome – hardly without precedence – was

---

3   Ryder 2001.
4   Channel Nine, *60 Minutes* 2018.
5   *UPC* 2018.

the arrest of thirteen of the group that came to the aid of the hens. They were charged with multiple offences, including animal cruelty,[6] and prevented from providing the hens with the medical and other care they needed, leaving them in the hands of the owner of the farm who had subjected them to such torture in the first place. Eventually – two months later, coinciding with the activists' first court hearing – the farmer was also charged with five counts of animal cruelty.

Along with physical suffering, which dutifully remains centre stage of the animal-protection movement, anthropogenic interference in both captive and free-living populations affects other aspects of nonhuman animals' subjectivity. From social to environmental privation and deprivation, the absence of species-specific normativity and, in captive conditions, the absence of individual and community autonomy may contribute, as evidenced in Chapter 2 on attachment and Chapter 4 on spirituality, to the exacerbation of the already immense physical pain and of the grief suffered by nonhuman animals – a grief that is also shared by those humans who have *seen* and cannot *un-see*.

This chapter explores the foundations which predict that grief 'at a distance' – grief for subjects we do not personally know such as the billions of nonhuman animals suffering and dying in the Anthropocene – is possible and bears comparable emotional charge to the grief experienced in relation to the loss of proximal significant others; that this grief is legitimate based upon the violence that commodification imposes upon nonhuman animal subjects; and that the outer expression of such grief is necessary as it bears witness to both the reality of the human mourner's internal turmoil as well as the conceived reality of the mourned nonhuman animals. The chapter concludes with a brief exploration into the world of another potential victim of the nonhuman animal holocaust – the direct human perpetrator.

---

6   Osborne 2018.

Bearing witness: from open rescue to open mourning

Sit-ins and lock-downs[7]

On 18 February 1997 the ACT Magistrates Court at Canberra, Australia, handed down a landmark decision – or so it was believed at the time – on animal-rights activists' unsolicited visits to factory farms. In the early hours of 20 October 1995, a group of about thirty activists from Animal Liberation gained access to shed number five (out of seven) on Parkwood Eggs' premises in Belconnen, where some 260,000 birds were trapped in the 'battery system'. Photographic and video evidence was obtained, dead birds were removed and sick ones taken to the vet for treatment. The remaining ten activists chained themselves to the cages, alerting the police and media. Four of the activists, namely Patty Mark, Lynda Stoner, Margaret Setter and Mark Pearson, refused to leave, demanding their complaints against Parkwood Eggs be investigated. Their subsequent arrest led to the above mentioned court case.

While events that the organisers and participants explicitly refer to as *vigils* are a recent development in animal-rights activism in Australia and elsewhere, other forms of bearing witness have been in practice for decades. Visiting a farm, a slaughterhouse, sale yards and similar establishments enables activists to gather evidence of the various aspects of the exploitation system, which the industry tries to hide from public view and awareness. It is also an opportunity to rescue animals, particularly those whose condition is most critical: 'When we enter the dark and cruel underworld of animal agriculture, we immediately reach out for the most vulnerable individuals, the injured and sick ones,' notes Patty Mark, founder of 'Open Rescue'.[8] Open Rescue is an act of nonviolent civil disobedience aimed at helping innocent individual victims of unjust laws and at bearing witness to the ghastly conditions they are forced to live and die in. 'Open Rescue is about opening – a gate, a door, a cage, one's heart, mindset, identity,'[9] and the rescued

---

7   The terms describe activities of bearing witness with activists sitting with the captive animals or locking themselves to the cages and other infrastructure.
8   Mark 2014, 110.
9   Mark 2014, 110.

animals, nurtured back to health if lucky, become 'ambassadors'[10] for all other animals still trapped in the machinery of death, voices for that invisible 'face on your plate'.[11]

A more intimate and vigil-like aspect of these operations manifests when activists lock themselves to the cages or pens. Lynda Stoner explains:

> Locking down was also an exercise in solidarity with the animals we were trying to help. To be locked on and unable to go to the toilet or other is uncomfortable but we all knew there would be a point in the day when we could walk away […] Also, being chained to a structure on the one hand meant you felt more vulnerable because you would be slowed down should workers become aggressive but it felt right and solid and made us as a group more impenetrable.[12]

These actions are of a strictly nonviolent nature with activists often wearing signs stating that no harm is meant to person or damage to property, and biosecurity measures (for example, wearing biosecurity suits and disinfecting feet) are adopted to prevent exogenous contamination. Peaceful resistance to leaving the premises upon police request when activists' demands fail to be met often leads to arrests. Arrests, however, particularly when followed by trials, attract more media attention, and the plight of the activists and that of the nonhuman animals the activists represent remains in the public eye for longer periods of time. As a consequence, Stoner observed, police appeared to have been encouraged to not arrest, and went to great pains to avoid it.[13] Indeed, had the pig farmer not pressed charges against Anita Krajnc,

---

10 Donaldson and Kymlicka (2015) problematise the concept 'ambassador' within the 'refuge + advocacy' sanctuary model. I do not have a problem with the term itself, though I fully agree that sanctuaries need to move towards the 'intentional community' model. See also footnote 62 of Chapter 1.
11 Masson 2009.
12 Lynda Stoner, email, 2018.
13 This attitude appears to have changed lately. Proposals for the introduction of various forms of so-called 'ag-gag' (Bittman 2011) laws have been advanced, and in certain cases passed, in the US and Australia. Ag-gag laws effectively make it illegal to document and spread evidence of corporate animal cruelty

the founder of the vigil-oriented Save movement, in 2015, the vigil movement may have never grown to its current international extent.

In the case of the Animal Liberation activists, charged with trespass of Parkwood Eggs' premises at Belconnen, the charges were dismissed by Magistrate Michael Ward based on the presence of a 'reasonable excuse' for their action. Unlike 'lawful purpose' and 'lawful excuse', Magistrate Ward explained, reasonable excuse may lead to an unlawful action but the action may nonetheless be reasonable; that is, it would be accepted as such by a reasonable person.[14] For many years the defendants had worked hard on bringing the inherent cruelty of the battery system to the attention of the public and politicians, both state and federal, in an attempt to ban this system. A veil of bitterness descends upon Mark's face as she remembers telling a concerned audience in 1978 upon discovery of the existence of the battery system that they would have to be patient as it may take two years to ban the cage.[15] They believed that showing them what was truly happening would motivate action for change. Over the years, however, the public has proven more tolerant of animal suffering (possibly due to the fact that they can choose to not expose themselves to it directly) than it was initially imagined, and to date little has changed to ease animals' pain in caged or other settings. Given that the animal-activist defendants in the Parkwood Eggs' case had in the past adopted and exhausted more conventional means of persuasion, Magistrate Ward acknowledged that it was hard to see what else they could have done other than taking the action they did. Following the review of evidence obtained by the activists at the farm and subsequent veterinary reports of the rescued and dead hens, the magistrate made several observations and conclusions that both absolved the activists and portrayed a faithful picture of the battery-egg system itself and the system that is supposed to work towards prevention of cruelty but does not (for example, police

---

and abuse even when the abuse is in breach of the current welfare guidelines regulating the industry. Essentially, it is not the abuser that would be penalised and/or the evidence used to take appropriate steps in order to prevent abuse in the future; instead the governments will shoot the messenger.

14   Ward 1997, 16.
15   Mark 2014, 108.

and the RSPCA). The magistrate found that the defendants had a reasonable excuse and the evidence proved them right, 'their trespass caused no harm to anyone, no drop in production, no interference with work, no damage, and they did some positive good by helping obviously sick hens'.[16] Two veterinarians with little expertise on battery operations, hired by the RSPCA, examined the factory and found no problems. Magistrate Ward attributed a 'farcical nature' to this examination due to its brevity and superficiality – the vets took between one and one and a half hours to inspect the entire operation, which amounts to each vet examining well over a thousand birds per minute.[17] Nevertheless, the reports were satisfactory enough for the police to choose not to proceed with an investigation of Parkwood Eggs.

Another issue raised by Magistrate Ward is the incompatibility between the code of practice for 'animal agriculture' and operations like battery-egg production, in which many critical behavioural and physiological needs specified in the code are not met. The magistrate asks:

> How can the code demand, on the one hand, as a basic requirement for the welfare of poultry, a system appropriate to the hen's physiological and behavioural needs, and on the other hand, allow for a system inimical to those needs?[18]

The answer, while disturbing, is simple: the code provides guidelines, but they are recommendations only, not statutory. If legislation truly protected all animals from cruelty, 'we wouldn't have animal farms or slaughterhouses'.[19] Technically, as the Victorian code of accepted farming practice for sheep states, the code 'is not prescriptive because good stock-handlers need to be flexible in their approach to caring for their animals'.[20] But ultimately nonhuman animals are kept in these establishments for human economic gain; by implication, when profit and the wellbeing of animals clash, profit will inevitably be prioritised.

---

16   Ward 1997, 5.
17   Ward 1997, 14.
18   Ward 1997, 11.
19   Mark 2014, 109.
20   2007, 1.

Since these animals are awarded higher legal protection as property than as sentient individuals, it is the activists who try to help them who are charged with criminal conduct more often than the farmers and meatworkers who abuse them.

Vigils

Krajnc's trial is a case in point of activists being the target of criminal charges rather than the farmers and workers who abuse animals. In December 2010, in Toronto, Canada, a group of activists, including Krajnc, began holding weekly vigils for pigs en route to slaughter. The group adopted the name Toronto Pig Save. This was followed shortly by the formation of two more groups with a focus on chickens and cows. These vigils were, and continue to be, inspired by words attributed to Leo Tolstoy and featuring as a motto on the movement's website:[21]

> When the suffering of another creature causes you to feel pain, do not submit to the initial desire to flee from the suffering one, but on the contrary, come closer, as close as you can to him who suffers, and try to help him.

Through bearing witness, vigils help raise awareness of the plight of those animals whose lives and deaths are perceived as being an exclusive function of human interest, as well as erect metaphorical glass walls around slaughterhouse areas. The vigil participants document the condition of the animals, and when possible they alleviate their pain, for instance by offering water to dehydrated animals on hot days. One such act of 'interfering with property' by giving an 'unknown' liquid to visibly distressed and panting pigs[22] on a truck about to enter the slaughterhouse premises in June 2015 induced the driver to confront the activists, followed by the pigs' 'owner' pressing charges against Krajnc. Much like in the Canberra case, this trial offered the

---

21 https://thesavemovement.org/.
22 Based on video evidence the veterinary witness estimated that the pigs were breathing 180 breaths per minute (reported in Craggs 2016), which is three to four times the normal rate.

opportunity to scrutinise animal-farming practices and slaughter. Animal-rights activists around the globe followed the trial closely as real-time updates were being posted on social media. It was difficult at times to avoid the feeling that it was the meat industry and not the activist that was on trial as witnesses from various fields of expertise testified about animal sentience, the impact of the industry on the environment and climate change and even about the benefits of a plant-based diet.[23] The prosecution failed to demonstrate that the farmer had, to borrow Magistrate Ward's wording, a reasonable excuse to believe that the 'unknown substance' was anything but water. Krajnc pointed out that the group had been giving water to pigs for three years in the presence of police, taking that as an endorsement; Justice David Harris noted that it was evident from the Toronto Pig Save's program that they were trying to help pigs not harm them, and ultimately those pigs were slaughtered and the operation effectively was not disrupted.[24] The charges were dismissed. While there were several Save groups already in existence at the time nationally and internationally, many more have sprung up over the following years as the movement continues to expand in number and focus.[25]

Public vigils are not new to the animal-liberation movement, including in Australia. Following similar events in Spain, Chile, Germany, France and Peru, in 2012 Animal Liberation Victoria, Australia, organised the first large-scale vigil for nonhuman animal victims of direct and indirect human violence. A hundred activists gathered in Melbourne's City Square, each holding an animal's cadaver – pigs, sheep, kangaroos, chickens, and others – collected from factory farms, paddocks, shelters, roads. The following year the vigil was repeated with 200 mourners. Standing in front of a tombstone, which read 'In Memory of the Unknown Animal', the mourners were brought the dead bodies, one by one, with a striking of a drum accompanying the respectful passage of each body.[26] 'The tribute has an emotional

---

23 Craggs 2016.
24 Craggs 2016; Craggs 2017.
25 Originally an 'animal save' movement, it has grown to include the 'climate save' and 'health save' movements, according to the website.
26 A moving video is available at: https://bit.ly/36KfMJw.

effect on passers-by,' writes Capps, 'to whom it seems to pose the question: which lives are grievable, and why? How do we decide that only some lives possess inherent worth?'[27]

Indeed, more than other forms of bearing witness, vigils appear to incorporate a delicate balance between exposing the pain of the nonhuman animal and that of the human animal who is incapable of turning away, of pretending she had not witnessed the suffering of the nonhumans or that their suffering does not matter. By validating the perceived inherent value of these animals' lives, the vigils also validate humans' grief for these lives, often considered as an emotional overreaction or even as plainly ridiculous. Ultimately, vigils may be viewed as a perhaps inevitable development within the new wave of the animal-protection movement, which, in Australia, began in the 1970s[28] with attempts to improve farm animals' wellbeing, for instance by campaigning to eliminate certain practices such as the battery-egg system and the mulesing of lambs.[29] With greater insight into the reality behind the closed doors of the 'animal agriculture' establishment came greater awareness of its toll on nonhuman animals themselves, but also on humans and the socio-natural environment at large. This coincided with human emotion (traditionally held in lower regard compared to 'reason') beginning to lose its stigma as more scrupulous research into emotion began in the 1990s, as discussed earlier, showing not only that emotions can aid the reasoning process, as opposed to disrupting it as previously believed, but that emotions are essential to it.[30]

Simultaneously, in the past two decades, refined research methods and a shift in attentional focus have enabled a more integrated view of nonhuman animals, progressively dismantling the notion of nonhuman animals as instinctual and behavioural automata by revealing their sophisticated social, emotional and cognitive skills. These and other factors, such as the realisation that while animals remain a source of profit their wellbeing will always be at risk, have enabled the

---

27   Capps 2014, n.p.
28   For a more comprehensive presentation of the period between 1970 and 2015 see Villanueva 2018.
29   e.g. Mark 2014; Townend 2017.
30   Damasio 2006 [1994].

animal-rights movement to evolve from its beginnings that focused largely on regulation into an abolitionist program of social justice.

While prejudice against other animals persists and animal activism continues to draw accusations of sentimentality, participants in public vigils for nonhuman animals are an embodiment of this program, whether they are addressing consumers of animal products in city squares or interacting directly with farmers, truckers and meatworkers in the field. The latter, much like the nonhuman animals who suffer at their hands, are themselves to various extents victims of society's demand for animal flesh and other products, an issue which will be considered in the last section of this chapter.

Nevertheless, like any other initiative, the vigil initiative, particularly some of the approaches by the Save movement, is vulnerable to criticism. Some of this criticism is of a more general nature with critics seeing the vigils as human-centred, serving human purposes only and lacking the capacity to help nonhuman animals. Other concerns are about more specific issues. For instance, the transport truck, as Melissa Boyde,[31] scholar and advocate who lives with various rescued farm-animal species, aptly points out, is in itself a stressful environment: animals have restricted movement, and while in transit they need to balance to stay upright to avoid injuries of various kinds. There is no provision in transit for food or water, whatever the conditions. Further, most animals on the way to slaughter would not have come from a loving environment that inspired trust in humans. The animals may have been handled roughly while they were being moved onto the truck, which is certainly always the case for hens, but also other animals, especially if they try to resist. The animals may also have been moved by truck several times: from the farm to the sale yards, from the sale yards possibly to another farm or a feedlot before again being loaded onto the truck taking them to their death. As a consequence, and while it is far from the activists' intent, the presence of the activists, especially when they stand in close proximity to the truck, may stress these animals further as they may not in such a short time be able to come to differentiate between the well-meaning mourning humans and the humans they have experienced before.

---

31   Melissa Boyde, email, 2018.

When activists bring along their dogs and even let them stand with their front legs leaning against the truck (a posture mimicking that of the mourning humans), which occurred on at least one documented occasion in Spain, the result will most likely be a substantial increase in the stress levels of the animals in the truck.

To avoid traumatising the undoubtedly already highly stressed animals even further, some of the basic requirements for road vigils should include maintaining respectful distance to the truck (perhaps with the exception of when animals are being offered water), silence, no abrupt movements, and neutral clothing with no bright colours or vivid patterns, which usually scare animals as discussed earlier. There may be other issues the movement needs to consider. Addressing constructive criticism and making space for changes and improvements are paramount for the long-term health of any movement, and critical especially for activities that inevitably involve speaking for and representing nonhuman animals. Nevertheless, vigils in general remain a powerful conceptual and potentially political tool within the broader movement for the liberation of nonhuman animals and cessation of their relentless suffering, physical and psychological, at the hands of humans.[32]

Grievability of unknown animals

Grief is a deeply intimate affair, affecting a range of psychobiological levels and exhibiting the potential for long-term compromising effects on the wellbeing of the affected organism, including threats to its mere existence. Simultaneously, in socio-political settings the outer expression of loss, or lack thereof, is almost by definition a political act, as it defies, or supports, the selective social normative with its tendency to dictate who is worthy of mourning and hence of life itself.[33] The latter is particularly reinforced in relation to the lives and deaths of nonhuman animals, and is reflected in the widespread attitude towards human mourners of nonhuman animals, whose grief is usually not recognised, not admitted,

---

32 Dr Alex Lockwood (2018) theorises on the topic based on his own involvement in the vigil movement.
33 Butler 2004; Stanescu 2012.

and oftentimes even derided. James Stanescu pertinently employs Judith Butler's considerations of grievability, originally developed within the queer context, and extends the analysis to a cross-species context: inclusive of other animals, our relationships with them and feelings for them.[34] Butler wrote: 'To the extent that homosexual attachments remain unacknowledged within normative heterosexuality [...] they will not be an attachment that can be openly grieved.'[35] Similarly, humans' attachments to nonhuman animals are also taxed by socio-political determinants based on speciesism. The grief at the loss of a nonhuman animal falls under the category of disenfranchised grief, which describes socially unsanctioned bereavement, including losses that are not recognised as genuine losses.[36]

Human grief for nonhuman animal family members is rarely, if ever, considered to be on a par with the grief humans may experience at the loss of another human. This view is contrasted by recent findings, which demonstrate that for some humans nonhuman animals are perfectly adequate partners in attachment relationships, considered within the framework of attachment theory. Nonhuman animals are capable of providing their companion humans with both a safe haven in distress-inducing situations and a secure base for exploration and growth, two regulatory functions of attachment figures.[37] However, it is not insignificant, as Rockett and Carr point out, that such results, which become evident utilising physiological measures of distress and astute experimental designs, may be obfuscated if self-report methods are relied upon. In fact, as has been demonstrated, in some cases – and possibly due to defensive processes of the declarative domain – the consciously perceived and/or reported significant others of primary importance do not match the choice the subconscious makes.[38] Similarly, it also appears that when humans are allowed to discuss their attachments to nonhuman animals

---

34 Stanescu 2012.
35 Butler 1993, cited in Stanescu 2012, 568–9.
36 e.g. Cordaro 2012.
37 Zilcha-Mano, Mikulincer and Shaver 2012.
38 Rockett and Carr 2014.

anonymously, they are more likely to attribute higher emotional value and intensity to these attachments.[39]

The capacity to form close and unique relationships of mutual dependency with other animals has obvious implications for grief at the loss of such relationships. Jeffrey Masson's book *Lost Companions* is a powerful testimony to the beauty and depth of these relationships. But as Kelly Oliver points out, referring mainly to human–nonhuman companion-animal relationships: 'To love animals is to be soft, childlike, or pathological. To admit dependence on animals – particularly emotional and psychological dependence, as pet owners often do – is seen as a type of neurosis.'[40] Fearing stigmatisation, humans may be reluctant to openly admit such grief, as Millie Cordaro reminds us, warning of the disenfranchised griever's intensified vulnerability to unresolved grief and encouraging counsellors to adopt a non-judgmental, supportive attitude towards their clients.[41]

The spectre of speciesism and instrumentalisation further dictates, in relation to nonhuman animals of various species, the dynamics of love and loss to which a human is expected to adhere in order to comply with social standards of 'normality' and acceptable conduct. Therefore, while a certain level of attachment is condoned, in fact expected, when it comes to traditional companion-animal species such as dogs, it is generally perceived as absurd to protest, let alone grieve, the slaughter of the billions of equally sentient individuals pertaining to animal species traditionally usurped for 'meat' and other commodities.

However, a growing number of humans have moved away from the culturally determined ethics based on species segregation, and embraced and internalised the philosophy and praxis of interspecies equality. The empathic identification with the pain and suffering of other animals drives these humans into animal rescue and direct care, forms of advocacy that aim to educate about nonhuman animals' suffering and eradicate it, or a combination of both. As a consequence, these individuals are continuously exposing themselves as witnesses of direct or indirect violence, and as such become vulnerable to

---

39 Reported in Taylor and Fraser 2019.
40 Oliver 2011, n.p.
41 Cordaro 2012.

psycho-physical stress and trauma with potentially detrimental outcomes. It is elemental to recognise that for many of these individuals this 'lifestyle' is not a choice, but an imperative defined by their inability to *unsee*. This in turn defines who they are as individuals and informs an entire cognitive-emotional spectrum underpinning their identity. A pig, a cow, a goat, a sheep, a rabbit, and any other animal trapped in the system of (ab)use or victim of anthropogenic violence in the wild is not simply a number, an object, somebody's property. Every single one of them is a living, feeling being who deserves a species-specific life, free of preventable suffering. As such, their lives are precious and their deaths are grievable, just like everyone else's.

Public manifestations of open mourning thus validate the lives of the mourned animals as well as the lives of their conspecifics who continue to suffer and die under the violent pretence of normality. The vigils demonstrate that the lives of these animals are not a mere commodity as it is usually perceived, but that they are lives in the full meaning of the word, that they matter and are grievable. Simultaneously, the vigils validate the mourners themselves whose very being is in part determined by the sentiments and relations to these animals and who have found the strength to expose their vulnerability in a world where the latter is also regulated by socially determined modes and levels of affordability. Such public mourning may also offer a much-needed support to the numerous humans who themselves bleed, metaphorically speaking, from the uncountable deaths of their perceived nonhuman kin, but do so quietly, isolated, and perhaps isolating.

Despite its potent emotional charge comparable to that experienced in relation to other humans, the grief for nonhuman animals, particularly those animals the purposive and systematic exploitation and murder[42] of whom is socially acceptable and normalised, bears little social consent. Consequentially, many humans living with such grief do not only suffer the internal turmoils intrinsic to this feeling, but their grief, at the delicate time when a gentle, sympathetic hand may be most appreciated, is often met with indifference or even outright animosity by both their closer

---

42   In our current legal system the term 'murder' is used only in reference to the killing of other humans; with the ethical context based on interspecies equality the term is naturally extended to include lethal violence against all animal species.

social circles and the society at large. Social support, particularly having someone to confide in, someone whom a person can trust to understand and respect their feelings, can be beneficial in coping with grief in general.[43] Many humans grieving for nonhuman animals lack such support. A person who suffers profoundly from such grief is often subjected to further violence in the form of negation and derision of their feelings by others. To become part of and help build a mutually supportive and caring community of individuals affected by such grief both strengthens the grieving person and opens doors for others who may be suffering from similar ailments yet have not been able to express and hence appropriately address their suffering due to an aversive social context.

Vicarious trauma

Professionals and volunteers working with victims of violence and ill-fated events and circumstances are vulnerable to *vicarious traumatisation*, a term used by I. Lisa McCann and Laurie Anne Perlman in 1990 to refer to potentially harmful outcomes for therapists following exposure to their clients' traumatic material. Vicarious trauma in essence describes traumatisation induced by exposure to a primary victim's first-hand experience, manifesting symptoms mirroring those experienced by the primary victim. Like primary traumatisation, vicarious traumatisation may have long-term consequences for the vicarious sufferers' personal and professional lives and can negatively impact victims in their care. Carers may succumb to compassion fatigue and burnout as a result of ongoing exposure to emotionally challenging situations, particularly when such are accompanied by the carers' sense of ineffectiveness in helping and protecting the victims.[44] As such, the phenomenon of vicarious trauma has received increased attention in professional domains most at risk of developing such trauma within the human context – inclusive of but not limited to health and mental-health professions. This attention aims at spreading awareness both of the risk and of diversified strategies, centred on self-care, promoting vicarious resilience and trauma

---

43   Parkes 2009 [2006], 186, 188.
44   Berthold 2011.

stewardship. These strategies include, as helpfully summarised by S. Megan Berthold, the need to take time off, to exercise regularly, to follow a healthy diet, to get enough sleep, and similar habits that help preserve or develop good mental and physical health.[45]

Humans working with nonhuman animal victims, both in direct care and in advocacy, are equally vulnerable to vicarious trauma and burnout. Recent studies that have looked at the effects of working with victims of violence in sanctuary and shelter workers and the emerging evidence of psychological distress in veterinarians and 'animal control' workers further testify to the reality of the impact the suffering of nonhuman animals can have on those human animals who are left to deal with the mostly human-induced damage, and the need for strategies to be adopted by workers and activists to protect themselves from such and promote the strength needed to continue the work.[46] However, their vulnerability is not always appreciated by the broader society and oftentimes even by the groups and organisations'they work with. As Bradshaw, Borchers and Muller-Paisner point out, this results in nonhuman animal care providers and advocates receiving far less moral support and – partly due to financial constraints – far less training than their human-oriented colleagues. They also remind us that nonhuman animals' legal inequality – for example, animals being considered as property – may represent an additional impediment to helping the victim and preventing further abuse, as seen in the case of the New South Wales hens from the introduction to this chapter and other listed examples. While helping a human in need is condoned, in fact expected, helping a nonhuman animal in a similar situation may not only be met with lack of understanding from the rest of the human community, but may also incur legal sanctions, as the above authors note. However, awareness of human vulnerability in animal advocacy and care seems to be growing as symptoms become harder to neglect. pattrice jones's book *Aftershock*[47] has become an important tool for understanding and coping with such

---

45 Berthold 2011.
46 See, respectively, Bradshaw, Borchers and Muller-Paisner 2012; Nett et al. 2015; Tiesman et al. 2015.
47 jones 2007.

traumatisation for many animal activists involved in different advocacy and care activities around the world.

Vicarious loss

The constellation of potential vicarious affects based on empathy towards and identification with other subjects includes vicarious loss. Vicarious loss was termed 'vicarious grief' by Robert Kastenbaum, who described it as 'the sorrow one feels for a loss suffered by another person'.[48] Vicarious grief, Kastenbaum explains, can induce affective states and responses comparable to those suffered by the individual experiencing direct personal loss, including weeping, a feeling of emptiness and constriction, difficulties sleeping, and loss of appetite. The concept of vicarious loss is later elaborated by Therese A. Rando, who terms it 'vicarious bereavement' and defines it as 'the experience of loss and consequent grief and mourning that occurs following the deaths of others *not* personally known by the mourner'.[49] Rando proposes the existence of two types of vicarious bereavement. Type-I involves losses that are solely vicarious, that is, upon empathic identification with the primary mourner, the vicarious sufferer feels what the grief must be like for the primary mourner. Type-II includes Type-I bereavement plus an added personal component, reflecting either or both an exacerbation of the vicarious mourner's reaction to direct personal losses, and a perceived violation of one's assumptive world. Like vicarious trauma, vicarious grief can affect various professionals[50] and volunteers working with human victims as well as those working with and for nonhuman animal victims.

Vicarious grief, as the term implies, presupposes an intermediary subject – a primary mourner – who is directly affected by the loss and whose loss and its acuteness are then absorbed by another person – the secondary mourner. In many cases it may be this kind of vicariousness that induces humans' grief for unknown nonhuman animals. For

---

48 Kastenbaum 1987, 447.
49 Rando 1997, 259.
50 Dahlitz (2014) offers a brief but informative introduction to vicarious loss among paramedics.

instance, we may identify and suffer with the mothers and infants who are subjected to systematic separation and bond-breaking in various exploitative settings within the food-production industry. We may identify and grieve with the sow who has accidentally, due to the lack of adequate living space, crushed her piglet to death. Bearing in mind the attention many mothers devote to pre- and post-natal activities and caring for the young – including nest building in the case of pigs, among others – there is scarce reason to believe that the sow remains emotionally unaffected when her infants die or are taken away from her. Even if we allow the possibility of emotional and mental lethargy developed by the sows under such physically and psychologically oppressive conditions, the sow will never habituate to the regime to the extent of turning into an inanimate machine; her psyche is a living organism and will remain so until her last breath. Until then, each instance of violence (or gentleness) will feed this complex apparatus, which humans share with other animal species.

Similarly, we may identify with and feel the pain of the animals in the wild who have lost their children, their companions, their community with its safety-promoting dynamics, and their habitats – homes – due to hunting, misinformed conservation attempts based on killing, human material expansion and other forms of direct or indirect anthropogenic violence. We may also deeply grieve when species face extinction, be it imminently physical extinction or psychological extinction and the change of a species' internal normative. In either case, multiple individuals must suffer, and in the case of physical extinction, die, for extinction to eventuate.

The nature of the pain some humans feel in the face of the above and similar (mostly anthropogenically induced) violence cannot be reduced to mere sorrow as expressed by others who upon learning about such brutality may indicate their regret for the mass suffering and continue with their lives mostly undisrupted. The pain of the human mourners for unknown animals can be deeply felt and comparable to the pain experienced after a direct proximal loss. Such feelings are possible as they originate in and rest upon the secondary mourner's empathic recognition of relatedness to both conspecifics and members of other species. We never in fact mourn the unknown, we always mourn the known.

The increasingly more open admission of human–nonhuman-animal mental and emotional comparability, including strong evidence predictive of nonhuman animals' capacity to experience grief, substantiates the legitimacy of the empathic projection that gives rise to grief for unknown animals. The artificial barriers constructed in service of the ideology of human superiority are slowly subsiding as humans are beginning to rediscover a space for themselves in the *communion of subjects* in nature, from which they had self-exiled, causing much grief to others and possibly to themselves. While it would be naïve, in fact potentially damaging, to promote sameness both in relation to nonhuman animals and to other humans, the relationality and relation*ability* embodied in an inter-species space of being together and knowing and recognising each other confirms the existence of comparability beyond physical form. It is this subjective encounter with the other that informs one's understanding of and attitude towards oneself and the rest of the world. It also defines one's spectrum of attention, which in turn influences what we find.[51]

Let us consider sheep, for example, a staple of many humans' diets and apparel, an objectified and instrumentalised lump of wool and tissue to be mutilated, traded and eventually slaughtered for human interest. If one lives with sheep, as I do, in an intersubjective space aiming for equality rather than exploitation, the general instrumentalising attitude is irreconcilable with what I see. My vision – like anybody else's vision – might be blurred, of course, by the nature of attention I apply, but my relationship with the sheep is *real* as are my feelings for them, feelings which extend to all sheep and other animals whom I have never met in person but, by analogy, *know* nonetheless. No reductionist 'truth' can change this internal truth that brings both joy and pain. And finally, should I seriously consider worrying about my intellectual integrity in my perception of the sheep when such is challenged by a visitor who cannot even distinguish one sheep from another among the highly distinctive faces of the sheep in question, which is usually the case? This phenomenon – the (in)capacity to distinguish faces – which is known as 'cross-race effect' within the human context but is equally applicable to interspecies settings, as discussed in Chapter 1, does not

---

51 McGilchrist 2009, 29.

reveal much about sheep's subjectivity. However, it does remind us of the precariousness of detached agnosticism in regard to other animals' physical and mental experiences, when in practice such agnosticism hypocritically promotes objectification and further violence towards animals. While we live, we will inescapably continue to impact their lives. The cognitive deficiency some humans admit to possess in regard to the existence, or lack thereof, of nonhuman animals' subjectivity and a human-comparable psyche is far from the neutral 'objective' position that it is often claimed to be. Such 'objectivity' is a collectively agreed-upon position, promoting a particular kind of attention and stemming from the belief in human supremacy and a history of fear conditioning. In practice it directly supports exploitation of and cruelty to nonhuman animals, while indirectly supporting violence against those humans who identify and feel with these animals.

The line between Rando's two types of vicarious bereavement is often diffused for the human mourner of unknown animals. In many cases, however, the grief goes beyond vicariousness. The terrified look of an individual in a transport truck destined for slaughter is coming from the eyes of a brother. I feel his fear, I feel his pain. I grieve his death and the death of each individual whose terrified gaze had been hidden from me but which radiates from pieces of flesh on supermarket shelves, and other products of violence encountered daily. Grief may source from the unbearable lightness with which the ontological genocide of certain species is implemented with indelible repercussions for the individuals and the species at large. Finally, grief may also arise from the savage disruption of these lives' imminent potentiality – of the kind Michael Marder advocates for plants,[52] that is, the potential of a life to be, and to be in a certain way and for a certain time. The scale of the horror of anthropogenic violence may indeed represent the apical difference between the loss of a proximal significant other (human or nonhuman) and loss at a distance, and in the latter case the assumptive world is therefore always challenged.

---

52   Marder 2013a.

The assumptive world

The assumptive world, as the term coined by Colin Parkes suggests,[53] is an individual's blueprint of assumptions and expectations consciously and unconsciously constructed from experience, which informs the way we perceive the world and ourselves in it, and against which we measure and evaluate the various events and situations in life. It grounds us, orients us and guides our thoughts and actions.[54] According to Ronnie Janoff-Bulman, there are three fundamental assumptions underlying humans' worldview, namely: a) that the world is benevolent – generally speaking it is a good place and most people are kind; b) that the world is meaningful, things make sense and happen for a reason; and c) that the self is worthy – by and large individuals perceive themselves as good, capable and moral.[55]

When loss or other traumatic events occur – proximal or at some distance – one's assumptive world may be shattered. To promote personal healing and wellbeing a revision of the assumptive world and an adaptation to the new reality may be necessary. But unlike in the event of a loss of a proximal loved one where such revision is encouraged and may to various extents be possible, when it comes to grieving for unknown animal victims, the tension between the assumptive world and the reality of anthropogenic violence against nonhuman animals can never really be resolved. My assumptive world, in which nonhuman animals are subjects, individuals and considered worthy of equal ethical consideration compared to humans, causes my grief. The grief, in turn, sustains my assumptive world. I do, to a certain extent, retain the above core assumptions, and it is critical for me to do so as it enables my advocacy for nonhuman animals, which is largely based on the belief that my fellow humans are essentially kind and that my actions will eventually contribute to a better world for all animals.

At the same time, I am surrounded by concentration camps and the supposedly inherently kind humans around me make the existence of these camps possible. If I adapt my assumptive world to the reality of violence, which for the sake of my own wellbeing may be desirable,

---

53 Parkes 1971.
54 See, for example, Beder 2005 for a brief overview.
55 Reported in Beder 2005, 258.

I may be vulnerable to desensitising. Desensitisation may occur either way: I could desensitise in relation to the nonhuman animals that I am trying to protect, and attempt to internalise the widespread normalisation of the violence against them, at least to the extent of practising quiet tolerance towards other humans and their often quoted right of choice of food and lifestyle, and abstain from voicing constructive criticism. I could, on the other hand, desensitise in relation to the humans that I am trying to protect other animals from, abandon the notion of humans' kindness and withdraw from the human world entirely, which would likely entail giving up public animal advocacy as well. However, desensitisation is not a healthy coping mechanism; it is a breakdown.

Hillman writes: 'Of course I am in mourning for the land and water and my fellow beings. If this were not felt, I would be so defended and so in denial, so anaesthetized, I would be insane.'[56] Nevertheless, by and large, insanity – along with softness implying weakness – is not attributed to the 'anaesthetized' human; instead it is usually ascribed to those, including mourners of nonhuman animals, who are unable to make themselves perceive the rest of the world as an instrumentalised commodity. A case in point was Kelly Atlas's heartfelt public plea to consider the slaughter of 'food' animals for the violence it really is. After recounting the story of Snow, the disabled hen she had rescued from an egg-production facility, Atlas tearfully urged the audience to think of Snow when they see animal products on their plates, stating: 'It's not food, it's violence.' Atlas's protest attracted media attention, and Atlas quickly became subject to ridicule. 'Atlas' protest is indeed impassioned, and she does lose her composure,' Gazzola writes, adding:

> The problem with the [media] coverage is not the descriptions of her crying or even ranting, both of which are accurate enough. The problem is the implicit judgment that to care this deeply about Snow and other animals killed for food is some kind of psychosis.[57]

---

56 Hillman 1996, 39.
57 Gazzola 2014, n.p.

The weakness that motivates humans to conform to societal expectations on the other hand is perceived as sanity and strength. There is nothing insane, of course, about human and nonhuman animals enjoying relationships with others and forming groups with social norms and expectations; it is a psychobiological need enhancing physical protection and emotional support. What *is* 'insane', as far as the current Western-bound attitude is concerned, is to deny this innate vulnerability and attempt to disguise it even from ourselves by turning a simple and natural phenomenon like group formation, which is supposed to protect our fragile selves and promote strength, into a system so oppressive to nonhuman animals and so fragile in itself that its very survival relies on most humans' inability to look at what underlies it out of fear that, upon *seeing*, the self would spill like quicksilver. This is not strength.

R. Scott Sullender notes: 'Most of us, unless we are sociopaths, are touched by the tears and sufferings of other humans and even other living creatures. We are instinctually drawn to their aid.'[58] If we agree with Sullender, by allowing the society, of which we are agent constituents, to attempt to 'protect' our fragile selves by promoting safety based on disguise and denial of what a large majority may intrinsically perceive as ethically deeply compromised principles and practices (which is reflected, for example, in humans' resistance to witness procedures in slaughterhouses), we are not growing safer and stronger, but more fragile and more vulnerable, both as individuals and as a society, heading towards sociopathy and a devastating internal split. Simultaneously, the human victim-turned-perpetrator of such denial is perhaps indirectly but nevertheless actively participating in the violence inflicted upon both nonhuman animals and those fellow humans who are unable to adjust to the 'normalisation' of such violence. Finally, and importantly, the general denial is contributing also to the violence against slaughterhouse workers themselves, whose mental and physical health are heavily impacted by such work.

---

58   Sullender 2010, 193.

Farmers and meatworkers

Out of sight, out of mind

In the preface to the Russian edition of Howard Williams's book *The Ethics of Diet* is an account from Leo Tolstoy written in 1892:

> Once, when walking; from Moscow, I was offered a lift by some carters [...] On entering a village we saw a well-fed, naked, pink pig being dragged out of the first yard to be slaughtered. It squealed in a dreadful voice, resembling the shriek of a man. Just as we were passing, they began to kill it. A man gashed its throat with a knife. The pig squealed still more loudly and piercingly, broke away from the men, and ran off covered with blood. Being near-sighted, I did not see all the details. I saw only the human-looking pink body of the pig and heard its desperate squeal; but the carter saw all the details and watched closely. They caught the pig, knocked it down, and finished cutting its throat. When its squeals ceased the carter sighed heavily. 'Do men really not have to answer for such things?' he said.
>
> So strong is man's aversion to all killing. But by example, by encouraging greediness, by the assertion that God has allowed it, and, above all, by habit, people entirely lose this natural feeling.[59]

Tolstoy goes on to explain his attempt to convince his acquaintance to join him on a visit to the slaughter yards. The acquaintance, despite being himself a hunter and not unused to killing, was reluctant to accompany Tolstoy if slaughter was to be in progress at the time of the visit. To this Tolstoy aptly remarked: 'Why not? That's just what I want to see! If we eat flesh, it must be killed.' But the acquaintance insisted: 'No, no, I cannot!'[60] in an attempt perhaps to preserve Janoff-Bulman's third component of his assumptive world.

It is beyond the scope of this chapter to delve into the question of differences between hunting versus slaughterhouse and farm work.

---

59 Tolstoy 2009 [1892], 39.
60 Tolstoy 2009 [1892], 40.

Suffice to note that the physical distance between the hunter and the hunted, characteristic of most hunting practices, may allow an amplitude of psychological distancing that may not be affordable on the killing floor of a slaughterhouse, and many forms of hunting[61] may also be perceived as being more of a 'fair game' compared to the slaughterhouse into which nonhuman animals are forced in order to be slaughtered with no chance of escaping that fate.

Tolstoy, as we learn, ended up visiting the slaughterhouse and described his experience in some detail in the following paragraphs. He laments most humans' reluctance to face the reality surrounding the animal food they consume, but concludes on an optimistic note, pointing out that the vegetarian movement was spreading rapidly, publications on the topic were becoming more and more numerous and the number of vegetarian hotels and restaurants especially in Germany, England and the United States was increasing every year. It was perhaps not spreading as fast as Tolstoy would have desired, but as he writes, 'the moral progress of humanity – which is the foundation of every other kind of progress – is always slow'[62] and it is its consistency and uninterruptedness, as he saw it, that demonstrate that the progress is true progress, not only casual. The book *The Ethics of Diet*, which he wrote this preface for and which includes a wide range of thinkers from different backgrounds and eras, all promoting an ethical life and kindness to nonhuman animals, was supposedly proof of such consistency and true moral progress.

Meanwhile, in nineteenth-century Japan under the Meiji rule, a 1200-year-long ban on the consumption of 'meat' was lifted with the Emperor in 1872 (six years before *The Ethics of Diet* started to appear as a series of articles in the monthly journal of *The Vegetarian Society*) publicly announcing, as part of the attempt to Westernise the country, that he would celebrate the new year by consuming animal flesh.[63] This would have encouraged other Japanese people to also consume animal flesh and openly speak about it, something that had not been customary in the past.[64] That 'meat' was banned in Japan prior to the Meiji period

---

61 Canned hunting being one of the exceptions.
62 Tolstoy 2009 [1892], 46.
63 See, for example, Watanabe 2005; Krämer 2008; Mowat 2009.

appears to be a gross overstatement. Despite Emperor Tenmu's ban on the flesh of many land animals in 675 CE,[65] Japanese people of different social classes, regions and eras continued to consume it, albeit on a much smaller scale compared to Europe. Historically such flesh was never a central part of the Japanese diet, and systematic raising of animals for food had not developed until recently with some parts of the countryside not integrating flesh into their diet until the middle of the twentieth century.[66] A combination of the Buddhist notion of compassion for all life and the Shintō notions of purity underpinned the consumption of flesh as a taboo. While flesh continued to be eaten, it was not done so publicly, and various euphemisms were in use to describe different types of animal flesh and encourage the widespread cognitive dissonance; for example, boar flesh was known under the Japanese equivalent for peony, horse flesh was referred to as cherry blossom and the flesh from deer as maple.[67]

Naturally, for humans to be able to consume this flesh, someone had to kill the nonhuman animals and process the parts, which gave rise to the so-called Buraku communities, also known by their pejorative name Eta,[68] which according to the Buraku Liberation League website signifies 'extreme filth'. Highly vilified and ostracised, the members of these communities were considered immoral and impure for engaging in socially stigmatised professions, which included animal slaughter and processing (for example, tanning). These communities lived segregated on the periphery of towns and villages, eventually attracting other occupationally or socially marginalised individuals and groups, including beggars, criminals and entertainers.[69]

---

64 The Emperor's move towards looser dietary ethics was not exempt from criticism and resistance, including an alleged attempt to break into the Imperial Palace by ten monks, which left several of the latter injured or dead (Watanabe 2005).
65 According to Watanabe (2005, 3) the ban covered species that lived with humans (such as cows, chickens, horses, dogs) and monkeys, but it did not include wild birds and other wild animals.
66 Krämer 2008, 40.
67 Mowat 2009, 43.
68 Donoghue 1957.
69 Donoghue 1957, 1001.

Apart from working with 'dirty' cadavers, another factor that contributed to their stigmatisation was their practice of consuming 'meat' publicly – for centuries only members of these communities did so,[70] while the rest of the society engaged in such activities in private. Descendants of the Buraku people remain stigmatised to this day.[71]

When Commodore Perry and his crew reached Tokyo in 1853 on behalf of the US government in an attempt to open up trade between the two nations, they marvelled at the abundance of bird life fearlessly perching on the masts and landing on the decks and then began to shoot the birds. It may come as no surprise that the Japanese witnesses were outraged at the Americans' actions, reportedly exclaiming: 'What savages!'[72] While I have not been able to verify that this event occurred as an empirical fact, the perceptual truth of those bearing witness[73] to this event is far from negligible in a world where it is the perceptual truth that largely informs our being and our actions, as well as shaping our assumptive world. Perceptual truth can empower humans and their attempts for ethical progress, but it can also nurture the illusion of ethicality, based on denial, as was the case with historical Japan and remains the case today in relation to animal exploitation and slaughter.

Meanwhile, in 'Savageland' the tide was also turning but in the opposite direction, as already suggested by Tolstoy's account above. Australia was no exception. Edgar Crook provides a helpful overview of the vegetarian movement's unfolding in the nineteenth and beginning of the twentieth century.[74] The first vegetarian society in Australia was formed in 1886 in Melbourne. Vegetarian restaurants were also in

---

70 Krämer 2008, 37.
71 e.g. Sunda 2015.
72 Watanabe 2005, 7.
73 Psychoanalyst Dori Laub recounts a disagreement between historians and psychoanalysts over a recorded testimony for the Video Archive for Holocaust Testimonies, Yale University. The historians argued for the discrediting of the testimony in its entirety given that this witness failed to reproduce correctly known empirical facts regarding the event in question, but the psychoanalysts disagreed, pointing out that by insisting on empirical facts, the historians failed to notice the internal *truth*, in this case the 'very secret of survival and of resistance to extermination' (reported in Oliver 2001).
74 Crook 2014 [2008].

existence at the time as were schools founded by vegetarians and publications promoting the diet. While most vegetarians adhered to the diet for health reasons, Crook notes, many incorporated animal welfare into their activities, including opposition to vivisection and hunting. Consuming dairy and eggs was not considered problematic, with a few exceptions relying on plant-based food alone. One such exception was suffragist Henrietta Dugdale, who in 1883 also wrote a utopian futuristic novel depicting a society that is completely vegan.

Robert Jones was another early vegan in Australia. His lifestyle was inspired by his abhorrence of animal suffering, and his speeches were published both in Australia and England. Humans should, Jones urged:

> [c]ease their consumption of that grossest of all foods, dead flesh [...] The suffering of gentle, domestic animals by land and sea, in railway trucks and cattle-steamers, from thirst , hunger, cold, heat, overcrowding, fatigue, blows, terror, and sickness, not to mention their death-agonies, and the unspeakable horrors of the slaughterhouses are such as no pen can describe ...[75]

An anonymous neighbour of the London Smithfield meat market, in a letter penned on 20 December 1848 and published in the *Farmers Magazine* in early 1849, also describes abuses witnessed, namely:

> the incessant barking of dogs, the bellowing of the oxen and calves, the bleating of sheep, the grunting of swine, the roaring and swearing of men, with torches, passing to and fro among the frightened animals, and the continued sound of blows inflicted on the horns, heads, and bodies of the poor animals, produce an impression on the beholders that no person can adequately describe, and must be seen to be believed.

However, instead of calling on his fellow humans to adopt a fleshless diet, he calls upon the authorities to 'put a stop to one of the most direful nuisances in this metropolis'.

---

75   Jones 1888, reported in Crook 2014, 129.

In eighteenth-century England, in fact, social reformers began to argue for the replacement of private structures used for animal slaughter for human consumption (for example, butchers' sheds) with public slaughterhouses outside of the city limits. The benefits of the latter would have been the removal of animal slaughter from public view, an opportunity for higher hygiene standards, easier monitoring, and, according to the reformers, state regulation of the 'morally dangerous' activity of animal slaughter.[76] The presence of meat markets in the cities was causing increased unease among city dwellers and shopkeepers, calling for their translocation to the outskirts. The meat markets were perceived as encouraging primarily 'gin shops and public houses', as one interviewee reportedly put it, and animal slaughter 'educate[d] the men in the practice of violence and cruelty, so that they seem[ed] to have no restraint on the use of it'.[77]

Progressively, though not as a uniform process, slaughter around Europe as well as in the US became a concentrated operation, removed from public view. We can imagine how this has the capacity to engender a false feeling of purity and ethicality reminiscent of that from the pre-Meiji period in Japan among the general population who support animal slaughter and the existence of slaughterhouses by purchasing the products. Modernising farming and slaughtering techniques has made animal flesh more affordable and has increased the demand for the flesh. The increased demand in turn encourages further modernising resulting in even more affordable flesh and further removal of these animals from public view; all this at the expense of both nonhuman and human animal wellbeing. Free of the philosophical and cultural charge imposed by tradition in Japan, the slaughter of animals could nevertheless be seen as a taboo in the contemporary Westernised world, or as Georges Bataille put it: 'The slaughterhouse is cursed and quarantined like a boat carrying cholera.'[78] Everybody knows that slaughterhouses exist and that meat, leather and other animal-based commodities available for purchase are a product of these operations, yet few are willing to discuss it, let alone partake in witnessing the procedures. As the Smithfield market neighbour pointed

---

76 Fitzgerald 2010, 60.
77 Fitzgerald 2010, 60, citing Philo 1998.
78 Bataille 1997 [1929], 22.

out, the absence of direct witnessing makes it hard to capture the reality of it, hence leaving plenty of space for denial. Timothy Pachirat, the author of *Every Twelve Seconds*, warns in an interview with *The Atlantic* that there is much more to witnessing than the visual act, which may be easier to mediate compared to other senses, particularly the smell and the sounds of the killing floor. He would, at times, find himself marvelling at the aesthetic effects produced by 'the shades of red, purple, and green against silver gleaming metal'. The smell and the sounds, on the other hand, were much harder to negotiate; they penetrated his clothes, his skin, his mind, and stayed there. Concurring with the Smithfield inhabitant, Pachirat too laments the futility of using verbal – and even visual – images to present the reality of the slaughterhouse, a reality experienced daily through all their senses by the killed nonhuman animals and their human killers.[79]

The human toll

The intentional pursuit of ignorance exhibited by most consumers of animal flesh cannot alleviate an individual's (or a society's) responsibility for what is experienced at the sequestered contemporary *smithfields*, or, one may be tempted to say, in our contemporary 'Buraku communities'. Slaughterhouse work is not culturally stigmatised hence the so-called meatworker (slaughterhouse employee) is not discriminated against. However, the current demand for meat has resulted in increasingly larger slaughterhouses, situated in increasingly more remote locations, breeding unhealthy communities exhibiting a range of social and other problems, which are not found in communities with other 'manufacturing' industries.[80] Of particular concern is the increasing evidence of elevated crime rates in slaughterhouse communities compared to other communities, particularly violent crimes and sex offences, including domestic violence and child molestation.[81]

The source of this alarming situation could be tracked to the fact that slaughterhouse work is physically demanding and dangerous, but

---

79   Pachirat 2011; 2012.
80   Fitzgerald 2010; Fitzgerald, Kalof and Dietz 2009.
81   Fitzgerald 2010.

also, and perhaps more importantly, psychologically taxing. Evidence has begun to emerge of the impact of such work on the workers' mental health, which induced Jennifer Dillard to advance the proposal for the recognition of such work as 'an ultra-hazardous activity for psychological wellbeing'.[82] Dillard introduces her essay on the psychological harm suffered by slaughterhouse workers with a quote by ex 'hog-sticker'[83] Ed Van Winkle, who used to work in a slaughterhouse in Iowa:

> The worst thing, worse than the physical danger, is the emotional toll … Pigs down on the kill floor have come up and nuzzled me like a puppy. Two minutes later I had to kill them – beat them to death with a pipe. I can't care.[84]

Virgil Butler, former worker in a chicken slaughterhouse, later turned anti-slaughter activist, documents his own experience of the slaughterhouse. He explains that most workers would use stimulants that helped them cope with the speed of the line as well as substances that help escape reality. One cannot talk about the emotional toll to co-workers, Butler laments, for fear of appearing soft, nor could one talk to friends and family since their inability to help made them uncomfortable. 'Out of desperation,' Butler continues, 'you send your mind elsewhere so that you don't end up like those guys who lose it,' providing two examples: 'the guy who fell on his knees praying to God for forgiveness' and another fellow worker who ended up in a mental-health institution for having nightmares of chickens coming after him. 'I've had those, too,' Butler adds, concluding:

> People tend to avoid you, even others at the plant, whether from instinct or because they know what you do and can't understand how you can do it night after night […] You feel isolated from society, not a part of it. Alone. You know you're different from

---

82   Dillard 2008, 407.
83   Worker who stabs hogs to make them bleed to death.
84   Dillard 2008, 391.

most people [...] They have not seen what you have seen. And they don't want to. They don't even want to hear about it.[85]

Dillard identifies two psychological frameworks underpinning slaughterhouse employees' psychological trauma, namely perpetration-induced traumatic stress (PITS) and doubling. Unlike vicarious trauma experienced by activists and carers where the secondary victim becomes involved in the aftermath of a traumatic event, or in any case had not themselves contributed to the trauma, PITS describes the phenomenon where traumatic stress occurs in a person because this person helped to create the traumatic situation for another subject. PITS is usually discussed within the human context, for example in relation to combat veterans, executioners and similar, but it is likely to affect slaughterhouse workers too, as they seem to exhibit symptoms relevant to this disorder.[86]

The phenomenon of doubling, on the other hand, was termed by Robert Jay Lifton[87] in reference to the Nazi doctors who effectively divided the self and constructed two separate functioning wholes, two selves that are capable of functioning as a whole self, depending on the situation. Hence the Nazi doctors retained the old, pre-concentration-camp self that enabled them to act as caring husbands, fathers and generally as moral individuals within their normative social circle, while simultaneously they were able to 'turn on' the other self when they were performing atrocious procedures on the human victims in the camps. A particularly poignant illustration of the power of doubling is Josef Mengele's rescue of a drowning Gypsy, whom Mengele after the saving act put on a truck heading towards the gas chambers.[88] Dillard sees – and I concur – instances of doubling within the slaughterhouse workers' population; for example in the case of Ed Van Winkle cited above, where Ed's 'normal' self recognises the sentience of the pig who nuzzles him 'like a puppy' while Ed's occupational self beats this sentient creature to death and *cannot care.*

---

85   Butler 2003, n.p.
86   Dillard 2008; MacNair 2002.
87   Lifton 1986.
88   Lifton 1986, cited in Bradshaw 2009, 202.

The extent to which these supposedly holistic selves can actually survive as separate entities in the long term is questionable, as testimonies of nightmares, anxiety and other side-effects on slaughterhouse employees (and their communities) indicate.

It is critical to note that even if slaughterhouse employees exhibited the utmost desire to treat the nonhuman animals with exceptional gentleness and care, this would not be possible in practice given the speed of the line and the need to reach the established daily quota for the sake of profit.[89] Other not directly anthropogenic factors within the slaughterhouse may cause great physical pain to nonhuman animals and potentially psychological harm to humans, such as malfunctioning stunning and cutting equipment, which can result, for example, in chickens being boiled alive. Slaughterhouse employees have to learn to 'deal with it' and not care because the line cannot be stopped. Like nonhuman animals, slaughterhouse workers also appear to be victims of the increasing demand for cheap meat; like nonhuman animals, they too appear to be largely voiceless, and the society at large bears this guilt.

Similarly to slaughterhouse employees, workers in large-scale industrial animal-agriculture businesses (where most of the animal products available on the market come from) are also subjected to psychological stress, deriving from the high-speed, physically demanding and potentially dangerous work[90] and from the violence resulting from the need to 'not care'. Animals in these intensive breeding and raising facilities are prevented from fulfilling most of their basic psychobiological needs by the nature of these establishments alone, whose sole purpose is the production of large quantities of meat and other products as fast and cheap as possible. The apparent desensitisation of the worker – this need to *not care* – may result in further unnecessary violence towards animals. The widely advertised alternatives to intensive factory farming, described as humane, organic, free-range, cage-free and other 'feel-good' terms, have not proven particularly humane either,[91] and in the end these animals, too, end up in the slaughterhouse. Patty Mark remembers:

---

89 See, for example, Compa's 2004 report for Human Rights Watch.
90 Gustafsson 1997.
91 For a more comprehensive presentation see http://www.humanemyth.org/.

The worst suffering and torment I've ever witnessed was in a New South Wales slaughterhouse when a group of free-range pigs were brought in for slaughter. Coming from their 'good life' on the paddocks, to the noisy, crowded kill lines where they could hear other pigs screaming, smell the blood; they panicked, anguished and in fear, their mouths foamed, their eyes rolled. No words can describe it.[92]

Further, even in truly small-scale operations, farmers employ psychological mechanisms to protect themselves from damage their occupation may imprint on their psyche, testifying perhaps to Tolstoy's (and many others') idea of killing – and perhaps instrumentalisation in general – being in discord with human nature. Farmers, for instance, may allow themselves to become attached to so-called 'breeding stock' but not to so-called 'store stock',[93] which may be interpreted as a form of doubling.

Returning to the vigils, in recognition of the possibility of the workers themselves being unwitting participants in the system that oversees the brutal mass murder of nonhuman animals for human consumption, activists attending on-site vigils may carry signs conveying this idea. One such sign spotted at a vigil in early 2018 and published on social media read: 'No hate for truckers'. The purpose of the vigils is not to expose individual perpetrators, or to harass them in any way; rather, when the opportunity arises, activists engage in conversation with truckers and other workers. The latter are sometimes hostile, but many times they admit that the system is inherently cruel and unjust. Over time with repetitive visits to individual establishments, activists may forge a relationship with the slaughterhouse management and other workers.

Activist and photographer Veronica Rios details an occasion when activists were granted access to the kill floor, and another when they were given the opportunity to rescue a hen.[94] In this emotionally charged event Rios could not choose among the hens in the box which

92   Mark 2014, 109.
93   Richards, Signal and Taylor 2013.
94   https://www.facebook.com/TheForgottenPhotography.

one to save and which ones to leave, asking the manager to do it for her. The manager, on the other hand, when invited to name the hen he had chosen to liberate was unable to do so, asking Rios to do it for him.[95] Some groups have even reached an agreement with slaughterhouses ensuring that trucks stop for several minutes before the entrance so that activists can bear witness.[96]

Without causing major disruption to the operation, such vigils nevertheless challenge at a fundamental level our *meat culture*,[97] along with assumptions and denials that make such 'culture' possible and even thrive. Despite (or perhaps because of) its strictly nonviolent, kind, open and respectful approach, the movement has attracted the attention of authorities, from the Ontario Provincial Police officer from the hate-crime unit closely monitoring the mourning 'extremists', as the officer herself revealed addressing the audience at the Ontario Pork Congress annual meeting,[98] all the way to the UK National Counter Terrorism Police Operations Centre (NCTPOC) staff advising the Association of Independent Meat Suppliers on how to respond to the vigils, as a leaked email, in which NCTPOC invites the industry to attend an educational seminar on the topic, shows.[99]

Meanwhile, the *extreme* suffering of animals continues. 'One minute there were truck-loads of living, breathing animals. The next minute the trucks were empty. Every hour. On the hour. All day long,' recounts Kristy Alger from the Tasmania Save group in a public Facebook post on 3 February 2018, adding:

The whole system is so mechanical. There's an input, a process followed by an output. But caught up in the midst of these mechanisms are beings not so different from you and I [...] And they're made nothing more than a basic input/output equation, as though there is no blood in their veins, no beating heart in their chest. All that just becomes a mess that needs cleaning up.

---

95   Veronica Rios, personal conversation, 2017.
96   Silvennoinen 2017.
97   Potts 2016.
98   Anderson 2018.
99   Newkey-Burden 2017.

Knowing what happens behind closed doors and the inability to prevent it leaves indelible marks, with vicarious trauma, grief and compassion fatigue common outcomes of such insight.

* * *

Sometimes I think that just through existing
one creates damage and disappointment,
and that love keeps us open, that
long, beautiful wound
that one has the choice
to do as little as one can
but that even that will not change one's allotted portion
of sadness and destruction – that it will be there
whatever one does to avoid it
and that the challenge is not to escape it,
try as of course we must,
but to think of life differently, as something
other than how we thought it was
No matter how careful one attempts to be
there is always greater care to be taken
— David Brooks, 'Damage', *The Balcony*[100]

The implications of nonhuman animals' massive exploitation and murder for the nonhuman animal victims themselves, for the growing number of humans who identify with these animals and suffer traumatic stress and grief as a result, and, last but not least, for slaughterhouse workers and their communities who find themselves in a situation where they have *no* choice challenge substantially other humans' often invoked 'right' to choose what to eat and how to live. Considering the potentially detrimental psychophysical toll on the human groups discussed in this chapter, along with the impact of animal agriculture on unprivileged human populations and on the planet as a whole, the plight of nonhuman animals becomes a human-

---

100 Brooks 2008, 65.

rights issue as well. Deforestation, land degradation, high water usage and pollution, biodiversity loss, and other factors (including human health risks) associated with animal agriculture induced the UN to issue a report emphasising the benefits of a plant-based diet for human and environmental health.[101] The report identifies four countries that have included sustainability issues into the national dietary guidelines, namely, Brazil, Germany, Sweden and Qatar. Australian efforts to follow suit were sabotaged by a strong media campaign run by the food industry, farmers and fishers, leading the UN to use Australia and the US as examples of 'what happens when government support is lacking or inadequate'.[102] This echoes the government's resistance to address the live-animal export trade, mentioned at the beginning of this chapter, despite cumulative evidence of the inherently torturous conditions on the ships, and many other issues that impact upon the wellbeing of animals – human and other – and the rest of life.

Along with oppressive systems that tend to 'shoot the messenger' rather than tackle the acts of the perpetrator, as in the case of 'ag-gag' laws championed by powerful industry lobbies, and the wilful ignorance exhibited by a frightening majority that has effectively given up their right not to be a perpetrator[103] and finds comfort in system justification, there is another factor that contributes to the status quo, partly related to both of the above two points: individuals feel powerless to change the system. Even when they recognise the system as unjust they may choose not to act. 'What difference will it make if I change my ways? Most others won't and the situation will remain the same,' is something that individuals advocating for animal protection and many other just causes hear far too often, in the streets, at dinner tables, and elsewhere. Yet, in the end it does all narrow down to personal choice. It is indeed unlikely that any one of us is able to eradicate all suffering and wrongdoings from the world. However, choosing not to change what is in our capacity to change is already making a difference and it will continue to do so – for the worse.

101 Gonzalez-Fischer and Garnett 2016.
102 Gonzalez-Fischer and Garnett 2016, 5.
103 Boyer, Scotton and Wayne 2016.

Coda: The precarious way ahead

> *Every story can be told another way, often*
> *becoming radically different in the telling.*
> *There are always other stories that say 'no'*
> *to the one that only recently enthralled and*
> *convinced us ... In stories, as in life, things*
> *are not always what they seem.*
> — Graham Harvey, *Animism*, 2004[1]

In *Animism*, Harvey takes us on an explorative journey of 'new animism', a term he uses to describe 'worldviews and lifeways in which people seek to know how they might respectfully and properly engage with other persons', human and other-than-human persons, 'all of whom are worthy of respect'.[2] The story, or stories, consolidated through so-called modernity but whose genesis reaches back several thousand years to sacred texts and less sacred practices of dominion and control have left us with a devastated planet and with substantially disenfranchised relational – and, consequentially, intimate – selves. A new project, one based on respect for all perceivable existence, may indeed be needed, with some urgency, if we are to halt, and possibly

---

1   Harvey 2005, xiii.
2   Harvey 2005, xiv, xvii.

reverse, some of the damage the *animal rationale*, deeply intertwined with the *animal spiritualis*, has brought about in his pursuit of wisdom and the 'good life'. The immediacy of the current situation (animal – human and nonhuman – suffering, species extinction, planet degradation, among others) and the specificity of it (for instance, the unprecedented growth of the human population) pose the challenging question of whether, or to what extent, a master narrative from the past (animist or other) may aid in the process of transformation of the present to ensure a better future, or any future at all.

Harvey points out that animism does not require the adoption of a vegan diet, and that it is, in fact, 'often (perhaps even *most* often) found among hunters and fishers'.[3] If the modernist-turned-animist would adhere to this principle, one would wonder where they would source nonhuman animal flesh and secretions from. At least four possibilities come to mind: 1) conventional farms where most other humans obtain theirs from, but the new animist would need to elaborate a conceptual framework to make the practice/industry ethically (including environmentally) acceptable; 2) enslaving nonhuman animals themself, giving them a 'good' life, but that too would leave the nonhuman animals with little self-determination and choice over their destiny, substantially prejudicing relational egality; 3) hunting free-living animals, and fishing, in which case the new animist would have to come to terms with their own contribution to the already dwindling populations[4] as well as, like in other cases, with the issue of the emotional pain of the killed animal's family and friends; or 4) partaking in the emerging fad, replete with hierarchisation, of 'environmentally sustainable' flesh, namely that of so-called 'pest' species,[5] whereby the acquisition of sustenance aids eradicating a species that had been introduced (in many cases intentionally) but is no longer desirable. One

---

3   Harvey 2005, 116.
4   A recent study of the biomass distribution on Earth (measured in gigatons of carbon) showed that only about thirty per cent of all birds on the planet are wild birds, the rest being 'farmed' (mostly chickens), while humans (about thirty-six per cent) and so-called 'livestock' (fifty-nine per cent) dominate the mammalian world with only about five per cent of all mammals being wild (Bar-On, Phillips and Milo 2018).
5   Armstrong and Potts 2014.

such example is the Australian possum who was introduced to New Zealand for the fur trade but is now persecuted as a pest.

Clearly, worldviews and practices that promote continuation of consumption of nonhuman animals based on the idea (implied or explicitly expressed) of equality of the nonhuman world can be unsettling for those humans who have exposed themselves as witnesses to anthropogenic violence against other animals – violence committed for the purpose of obtaining comestibles and other reasons humans find to interfere with or terminate nonhuman lives.

This book began with an overview of various factors that have contributed to the consolidation of prejudice against other animal species, a prejudice that precludes appreciation of their subjectivity and their cognitive and relational skills. I noted that the tide is turning in both popular and academic discourse where more interesting questions are being asked about nonhuman animals and more creative ways are being constructed to answer them, bringing us closer, perhaps, to becoming 'smart enough to know how smart animals are'.[6] Even if we never reach such a level of sophistication, existing data testify to other animals' human-comparable affective capacities that we are ethically compelled to cease ignoring.

This book addressed specifically the experiences of grief and spirituality. In both cases the affective component and its potency become obfuscated by attributing to the human interpretative domain a role greater than it deserves. To experience loss deeply, it is the attachment relation that is of primary importance. Other animals are not only capable of attachment relations, such relations are vital to them, as they are to us. Furthermore, the developed attachment styles may inform the intensity of grief and the capacity to cope with it, with potentially detrimental outcomes for many animals, particularly those in captivity who experience highly non-normative and disrupted upbringings and subsequent lives. Similarly, to meaningfully engage in the dance with animacy, one does not need highly elaborated concepts of divinities, the sacred or other interpretative solutions that humans often associate with spirituality. The relational dance that emerges when the experiential consciousness responds to the touch of other animacy affirms its

---

6   de Waal 2016.

ontology and is experienced as meaningful without the need for cognitive elaboration pre- or post-factum to confirm the experience as meaningful. Both captivity (particularly in its intensive forms) and the degradation of the planet most likely, and in their own ways, significantly impact upon this propensity of animals to engage with the agencies in the environment and the healing properties of such engagement.

The biopsychosocial parallels between humans and other animals enable empathic identification with those animals, and the recognition that they not only suffer physical pain but are, through privation and deprivation, also vulnerable to human-comparable mental pain. Such identification may lead to vicarious trauma and grief in those who refuse to look away from the ubiquitous suffering of nonhuman animals due to anthropogenic interference. Such identification also elicits the need and the will to terminate such suffering, which results in rescue activities as well as various forms of advocacy aimed at raising awareness among the rest of the public about the suffering they are supporting by the choices they make, and how they can adopt measures to prevent the suffering. The psychological impact on those who are forced to the 'front lines' to enable a constant supply of animal flesh to the ever-growing human population and its greed for such flesh is a further indication of the pressing need for reparation and reform.

In recent years, the capacities of plants have received substantial attention, both scientific and in popular debates. Admittedly, the possibility of plant sentience is often brought up by sensationalists attempting to undermine animal-rights advocacy and relativise the ethical efforts of this important social-justice movement in the most abhorrent fashion, principally, it would seem, to excuse their own participation in nonhuman animal suffering and slaughter. However, not everyone who advocates for plants' moral consideration is necessarily promoting animal exploitation. Michael Marder's plant-rights advocacy[7] is sometimes perceived as undermining animal-rights advocacy, but essentially it is simply an attempt at deinstrumentalising, to various levels, life generally. Recent research in botany unveils sophisticated sensory capacities and behaviours of vegetable organisms reminiscent of those pertaining to the animal

---

7   e.g. Marder 2013a; Marder 2013b.

realm.[8] The absence of neurons and a brain in plants makes drawing parallels between plants and animals difficult. Understandably, such parallels would be particularly unwelcome by animal advocates who spend substantial amounts of energy and time trying to save nonhuman animals and bring their plight to human attention. Introducing plant sentience into the discourse could trivialise their efforts and jeopardise the movement as a whole. It could, but it does not need to. A refined sensitivity to the issue of instrumentalisation, and the promotion of respect for all life, can only strengthen a movement that seeks to end injustice, such as is the animal-rights movement, which necessarily includes the rights of the human animal, too. Sensationalists and trolls may indeed attempt to introduce anxiety and question the purpose of animal liberation due to emerging and possibly unsettling new data on plants. However, just as the recognition of nonhuman animals' sentience did not reverse the ethical unacceptability of abusing humans, there is no reason to consider the implications of plant life outside the botanical realm.

'But where do we stop?' asks the ethically concerned but sceptical listener. 'How are we to live[9] in a world of constant change and new data?'

Advocates for the moral consideration of plants, for instance, are not suggesting that we should stop eating plants, albeit people may be invited to consider the possibility of eliminating from their diets plants whose reproduction capacity is completely destroyed once the individual plant is consumed (for example, tubers as opposed to apples whose seeds may produce new life after the flesh has been consumed). As Stefano Mancuso points out, along with lacking irreplaceable organs, a plant 'has a modular design, so it can lose up to ninety per cent of its body without being killed. There's nothing like that in the animal world. It creates a resilience.'[10] Many animals are of course capable of various degrees of regeneration, including limbs (for example, zebrafish) and even heads (for example, hydra),[11] but this is

---

8   For a brief overview see Pollan 2013, and listen to the ABC program presented by Barclay 2019.
9   Singer 1993.
10   Cited in Pollan 2013.
11   e.g. Zhao, Qin and Fu 2016.

clearly not the case for a chicken's wing or a lamb's leg. Furthermore, plants may benefit from contact with animals for seed dispersal, which may eventuate, for example, through animal faeces after the fruit had been eaten, or certain seeds may attach to the animals' fur or feathers and get transferred to a different location that way.

The main purpose of plant-rights advocacy appears to be the recognition that plants are living beings with an interest in life and wellbeing (though not necessarily comparable to the subjectivity recognised in animals), and to allow plants the possibility of reaching their life's full potential without precluding their chances for further reproduction. Thus attribution of worthiness of moral consideration to plants would have implications for genetic modification of plant organisms, for growing monoculture crops, sterile fruits, the extensive use of pesticides and herbicides, among others.

Perhaps the purpose – and the answer to 'But where do we stop?' – is to never stop.

In this book I have attempted to tell the animal story the way intuition, science and experience currently outline it. This story contrasts substantially with the story based on hierarchies and division told by modernity. It is also a call to action. Other stories may arise in the future, informing action then.

Works cited

Abbott, Alison (2015). Animal behaviour: inside the cunning, caring and greedy minds of fish. *Nature* 521(7553): 412–414, https://go.nature.com/3nio014.

Abreu, M.S., A.J. Friend, K.A. Demin, T.G. Amstislavskaya, W. Bao, and A.V. Kalueff (2018). Zebrafish models: do we have valid paradigms for depression? *Journal of pharmacological toxicological methods* 94(2): 16–22.

Adams, Carol J. (2010). Why feminist-vegan now? *Feminism & psychology* 20(3): 302–317.

Agamben, Giorgio (2004). *The open: man and animal.* Stanford: Stanford University Press.

Ainsworth, Mary D. S. (1967). *Infancy in Uganda: infant care and the growth of love.* Baltimore: Johns Hopkins Press.

Ainsworth, Mary D.S. (1969). Object relations, dependency, and attachment: a theoretical review of the infant-mother relationships. *Child development* 40(4): 969–1025.

Ainsworth, Mary D.S., and John Bowlby (1991). An ethological approach to personality development. *American psychologist* 46(4): 333–341.

Ainsworth, Mary D.S., and S.M. Bell (1970). Attachment, exploration, and separation: illustrated by the behaviour of one-year-olds in a strange situation. *Child development* 41: 49–67.

Ainsworth, Mary D.S., S.M. Bell, and D.J. Stayton (1971). Individual differences in strange-situation behaviour of one-year-olds. In *The origins of human social relations.* H. Rudolph Schaffer, ed 17–58. London and New York: Academic Press.

Ainsworth, Mary D.S., Mary C. Blehar, Everett Waters, and Sally Wall (1978). *Patterns of attachment: a psychological study of the strange situation.* Hillsdale, NJ: Lawrence Erlbaum Associates, Inc.

Albrecht, Glenn (2005). 'Solastalgia' a new concept in health and identity. *PAN: philosophy, activism, nature* 3: 41–55.

Alderton, David (2011). *Animal grief: how animals mourn.* Dorchester: Hubble & Hattie.

Alkon, Abbey, W. Thomas Boyce, Torsten B. Neilands, and Brenda Eskenazi (2017). Children's autonomic nervous system reactivity moderates the relations between family adversity and sleep problems in Latino 5-year old in the CHAMACOS study. *Frontiers in public health* 5(155), doi: 10.3389/fpubh.2017.00155.

Alzahrani, A.D.M. (2012). Effects of depressions, stress and other factors on cradling bias in Saudi males and females. Ph.D. thesis, The University of Edinburgh.

Amarello, Melissa, Jeffrey J. Smith, and J. Slone (2011). Family values: rattlesnake parental care is more than just attendance. *Biology of the rattlesnakes symposium.* Tucson, Arizona. Poster from the presentation available online: https://www.snakes.ngo/ABSsm.pdf.

Anderson, James R. (2016). Comparative thanatology. *Current biology* 26: R553–R556.

Anderson, James R., Dora Biro, and Paul Pettitt (2018). Evolutionary thanatology. *Philosophical transactions of the Royal Society B* 373(1754): 20170262.

Anderson, James R., Paul Pettitt, and Dora Biro, eds., (2018). Evolutionary thanatology: impacts of the dead on the living in humans and other animals (theme issue). *Philosophical transactions of the Royal Society B* 373(1754).

Andics, Attila, Anna Gábor, Márta Gácsi, Tamás Faragó, Dora Szabó, and Ádam Miklósi (2016). Neural mechanisms for lexical processing in dogs. *Science* 353(6303): aaf3777.

Animals Australia (2015). Newborn lambs are freezing. *Animals Australia*, 15 July, https://bit.ly/3jAMqAL.

Archer, John (2001). Grief from an evolutionary perspective. In *Handbook of bereavement research: consequences, coping and care.* Margaret S. Stroebe, Robert O. Hansson, Wolfgang Stroebe and Henk Schut, eds., 263–283. Washington, DC: American Psychological Association.

Århem, Peter, and Hans Liljenström (2008). Beyond cognition – on consciousness transitions. In *Consciousness transitions: phylogenetic, ontogenetic and physiological aspects.* Hans Liljenström and Peter Århem, eds., 1–25. Amsterdam: Elsevier.

Works cited

Armstrong, Philip, and Annie Potts (2014). The emptiness of the wild. In *Routledge handbook of human-animal studies*. Garry Marvin and Susan Mchugh, eds., 168–181. London and New York: Routledge.

Arvin, Heidi Louise, and Adrian Harrison Arvin (2010). *The linear heritage of women*. Bloomington: iUniverse.

Bachelard, Gaston (1994). *The poetics of space*. Translated from the French original (1958) by Maria Jolas. Boston: Beacon Press.

Bacon, Ellis S., and Gordon M. Burghardt 1983 [1980]. Food preference testing of captive black bears. *Bears: their biology and management* 5: 102–105.

Bagot, Rosemary C., and Michael J. Meaney (2010). Epigenetics and the biological basis of gene x environment interactions. *Journal of the American academy of child & adolescent psychiatry* 49: 752–771.

Baird, Robin (2010). September 2010 whale sightings [brief report]. Orca network, http://www.orcanetwork.org/sightings/sept10.html.

Balcombe, Jonathan (2006). Laboratory environments and rodents' behavioural needs: a review. *Laboratory animals* 40: 217–235.

Balcombe, Jonathan (2009). Animal pleasure and its moral significance. *Applied animal behaviour science* 118(3-4): 208–216.

Balcombe, Jonathan (2010). *Second nature*. New York: Palgrave Macmillan.

Barclay, Paul (2019). Can trees talk and think? ABC, 27 March, https://ab.co/30xqIeS.

Bar-On, Yinon M., Rob Phillips, and Ron Milo (2018). The biomass distribution on Earth. *Proceedings of the National Academy of Sciences of the United States of America* 115(25): 6506–6511, https://doi.org/10.1073/pnas.1711842115.

Barrett, H. Clark, and Tanya Behne (2005). Children's understanding of death as the cessation of agency: a test using sleep versus death. *Cognition* 96(2): 93–108.

Bartle, Philip (1977). Urban migration and rural identity: an ethnography of a Kwawu community. Ph.D. thesis, Legon, University of Ghana.

Bataille, Georges (1997 [1929]). Slaughterhouse. In *Rethinking architecture: a reader in cultural theory*. Neil Leach, ed. 21–23. New York: Routledge.

Bataille, Georges (1989). *Theory of religion*. Translated from the French original (1973) by Robert Hurley. New York: Zone Books.

Battaglia, Marco (2015). Separation anxiety: at the neurobiological crossroads of adaptation and illness. *Dialogues in clinical neuroscience* 17(3): 277–285, https://bit.ly/3luNw1X.

BBC (1999). South Asia elephant dies of grief. *BBC News*, 6 May, https://bbc.in/2H50Hrn.

BBC (2003). Dinosaur family footprints found. *BBC News*, 2 December, https://bbc.in/3nyA65P.

Bearzi, Giovanni, Dan Kerem, Nathan B. Furey, Robert L. Pitman, Luke Rendell, and Randall R. Reeves (2018). Whale and dolphin behavioural responses to dead conspecifics. *Zoology* 128: 1–15.

Bechtel, William (2015). Circadian rhythms and mood disorders: are the phenomena and mechanisms causally related? *Frontiers in psychiatry* 6(118), https://doi.org/10.3389/fpsyt.2015.00118.

Beder, Joan (2005). Loss of the assumptive world: how we deal with death and loss. *Omega* 50(4): 255–265.

Bekoff, Marc (2009). Animal emotions, wild justice and why they matter: grieving magpies, a pissy baboon, and empathic elephants. *Emotion, space and society* 2(2): 82–85.

Bekoff, Marc, and Jessica Pierce (2009). *Wild justice: the moral lives of animals.* Chicago: Chicago University Press.

Bennett, Peter M., and Paul H. Harvey (1985). Brain size, development and metabolism in birds and mammals. *Journal of zoology* 207: 491–507.

Berman, Michael R. (2001). *Parenthood lost: healing the pain after miscarriage, stillbirth and infant death.* Westport, CT: Bergin & Garvey.

Berns, Gregory S., Andrew Brooks, and Mark Spivak (2013). Replicability and heterogeneity of awake unrestrained canine fMRI responses. *PLOS ONE* 9(5): e98421, https://doi.org/10.1371/journal.pone.0081698

Berry, Helen Louise, Kathryn Bowen, and Tord Kjellstrom (2010). Climate change and mental health: a causal pathways framework. *International journal of public health* 55(2): 123–132.

Berthold, S. Megan (2011). *Vicarious trauma and resilience.* Sacramento, CA: CME Resource.

Beston, Henry (1949 [1928]). *The outermost house.* New York: Henry Holt and Company, Inc.

Biro, Dora, Tatyana Humle, Kathelijne Koops, Claudia Sousa, Misato Hayashi, and Matsuzawa Tetsuro (2010). Chimpanzee mothers at Bossou, Guinea, carry the mummified remains of their dead infants. *Current biology* 20(8): R351–352.

Bittman, Mark (2011). Who protects the animals? *New York Times opinionator*, 26 April, https://nyti.ms/3lhsa7U.

Bloch, Maurice (1981). Tombs and states. In *Mortality and immortality: the anthropology and archaeology of death.* S.C. Humphreys and Helen King, eds., 137–147. London: Academic Press.

Bode, Nikolai W.F., A. Jamie Wood, and Daniel W. Franks (2011). The impact of social networks on animal collective motion. *Animal behaviour* 82(1): 29–38.

Boehnlein, James K. (1987). Clinical relevance of grief and mourning among Cambodian refugees. *Social science and medicine* 25(7): 765–772.

Bolhios, J. J., and R.C. Honey (1998). Imprinting, learning and development: from behaviours to brain and back. *Trends in neuroscience* 21: 306–311.

Boswell, John D., ed. (n.d.). We're all connected. *Symphony of science*, www.symphonyofscience.com; direct access to music video: www.youtube.com/watch?v=XGK84Poeynk

Bothwell, Robert K., John C. Brigham, and Roy S. Maplass (1989). Cross-racial identification. *Personality and social psychology bulletin* 15(1): 19–25.

Bowlby, John, and Colin M. Parkes (1970). Separation and loss within the family. In *The child in his family: international yearbook of child psychiatry and allied professions*. E. J. Anthony and C. Koupernik, eds., 197–216. New York: Wiley.

Bowlby, John (1958). The nature of the child's tie to his mother. In *International journal of psycho-analysis* 39: 350–373.

Bowlby, John (1982 [1969]). *Attachment and loss: volume 1: attachment*. New York: Basic Books.

Bowlby, John (1973). *Attachment and loss: volume II: separation anxiety and anger*. New York: Basic Books.

Bowlby, John (1980). *Attachment and loss: volume III: loss, sadness and depression*. New York: Basic Books.

Bowlby, John (1990). *Charles Darwin: A new life*. New York: W.W. Norton.

Boyer, Kurtis, Guy Scotton, and Katherine Wayne (2016). Beyond complicity and denial: nonhuman animal advocacy and the right to live justly. In *Intervention or protest: acting for nonhuman animals*. Gabriel Garmendia da Trindade and Andrew Woodhall, eds., 159–182. Wilmington: Vernon Press.

Bradshaw, G.A. (2005). Elephant trauma and recovery: from human violence to trans-species psychology. Ph.D. Pacifica Graduate Institute, Santa Barbara, USA.

Bradshaw, G.A., and Robert M. Sapolsky (2006). Mirror, mirror. *American scientist* 94: 487–489.

Bradshaw, G. A., and Allan N. Schore (2007). How elephants are opening doors: developmental neuroethology, attachment and social context. *Ethology* 113(5): 426–436.

Bradshaw, G.A., Theodora Capaldo, Lorin Lindner, and Gloria Grow (2008). Building an inner sanctuary: complex PTSD in chimpanzees. *Journal of trauma & dissociation* 9(1): 9–34.

Bradshaw, G.A. (2009). *Elephants on the edge: what animals teach us about humanity*. New Haven: Yale University Press.

Bradshaw, G.A., Theodora Capaldo, Lorin Linder, and Gloria Grow (2009). Developmental context effects on bicultural posttrauma self repair in chimpanzees. *Developmental psychology* 45(5): 1376–1388.

Bradshaw, G.A., Joseph P. Yenkosky, and Eileen McCarthy (2009). Avian affective dysregulation: psychiatric models and treatment for parrots in captivity. *Proceedings of the Association of avian veterinarians.* 28th annual conference, Minnesota.

Bradshaw, G.A., Jeffrey G. Borchers, and Vera Muller-Paisner (2012). *Caring for the caregiver: analysis and assessment of animal care professional and organizational wellbeing.* Jacksonville: The Kerulos Center.

Bradshaw, G.A. (2013). Living out of our minds. In *The rediscovery of the wild.* Peter H. Kahn and Patricia H. Hasbach, eds., 119–138. Cambridge, MA: MIT Press.

Bradshaw, G. A. (2014). Elephants, us, and other family: how attachment theory has started a cultural revolution. Paper presented at *Affect dysregulation and the healing of the self.* Annual interpersonal neurobiology conference, UCLA, March 14–16.

Bradshaw, G.A. (2017). *Carnivore minds: who these fearsome animals really are.* New Haven: Yale University Press.

Bratman, Gregory N., J. Paul Hamilton, and Gretchen C. Daily (2012). The impacts of nature experience on human cognitive function and mental health. *Annals of the New York Academy of Sciences* 1249: 118–136.

Bretherton, Inge (1992). The origins of attachment theory: John Bowlby and Mary Ainsworth. *Developmental psychology* 28: 759–775.

Brooks, David (2008). Damage. In *The balcony.* St Lucia: Queensland University Press.

Brooks, David (2018). Grieving Kangaroo I, II, III. In *Kangaroos – the 100 days project, Arcohab,* https://www.arcohab.org/a-100-days-of-kangaroo.

Brooks, David (2019). *The grass library.* Blackheath: Brandl & Schlesinger.

Brooks, David (2020). The grieving kangaroo photograph revisited. *Animal studies journal* 9(1): 201–215.

Brooks Pribac, Teja (2013). Animal Grief. *Animal studies journal* 2(2): 67–90, http://ro.uow.edu.au/asj/vol2/iss2/5/.

Brooks Pribac, Teya (2020). On letting people die: coronavirus, geronticide and other worldly matters. *Counterpoint navigating knowledge,* https://bit.ly/3nlPgvF.

Broom, Donald M., and Andrew F. Fraser (2011 [2007]). *Domestic animal behaviour and welfare.* 4th edition. Wallingford: CABI.

Broom, Donald M., and Ken Johnson (1993). *Stress and animal welfare.* London: Chapman & Hall.

Broom, Donald M. (2003). *The evolution of morality and religion.* Cambridge: Cambridge University Press.

Works cited

Brosnan, Sarah F., and Frans B.M. de Waal (2003). Monkeys reject unequal pay. *Nature* 425: 297–299.

Brown, Culum (2014). Are fish far more intelligent than we realize? Interviewed by Joseph Stromber. *Vox*, 4 August. https://bit.ly/3lrkIan.

Brulliard, Karin (2016). Here's how scientists got dogs to lie still in a brain scanner for eight minutes. *The Washington Post*, 31 August. https://wapo.st/3jwE1hT.

Bubl, E., E. Kern, D. Ebert, M. Bach, and L. Tebartz van Elst (2010). Seeing gray when feeling blue? Depression can be measured in the eye of the diseased. *Biological psychiatry* 68(2): 205–208. doi: 10.1016/j.biopsych.2010.02.009.

Bugos, Peter E., and Lorraine M. McCarthy (1984). Ayoreo infanticide: a case study. In *Infanticide: comparative and evolutionary perspectives*. Glen Hausfater and Sarah B. Hrdy, eds., 503–520. New York, NY: Aldine.

Buraku Liberation League http://www.bll.gr.jp/eng.html.

Burt, Jemima, and Paul Culliver (2020). Queensland's near century-old crocodile Buka dies in captivity, leaving his partner in mourning. ABC, 5 August. https://ab.co/2SungI0.

Buschmann, Jens-Uwe F., Martina Manns, and Onur Güntürkün (2006). 'Let there be light!' Pigeon eggs are regularly exposed to light during breeding.' *Behavioural processes* 73: 62–67.

Bush, George, Phan Luu, and Michael I. Posner (2000). Cognitive and emotional influences in anterior cingulate cortex.' *Trends in cognitive sciences* 4(6): 215–222.

Butler, Judith (1993). *Bodies that matter: on the discursive limits of sex*. New York: Routledge.

Butler, Judith (2004). *Precarious life: the powers of mourning and violence*. London: Verso.

Butler, Virgil (2003). Inside the mind of a killer. *The cyberactivist*, 31 August. https://bit.ly/2H3uVuQ.

Capps, Ashley (2014). Then I will hold you in my arms: a must see tribute to animal victims.' *Free from harm*, 6 January. https://bit.ly/3cY4bHP.

Cassidy, Jude (2016). The nature of the child's ties. In *Handbook of attachment: theory, research, and clinical applications*. Jude Cassidy and Phillip R. Shaver, eds., 3rd edition, 3–24. New York: The Guilford Press.

Catanese F., R.A. Distel, F.D. Provenza, and J.J. Villalba (2012). Early experience with diverse foods increases intake of nonfamiliar flavors and feeds in sheep. *Journal of animal science* 90(8): 2763–2773.

Center for Biological Diversity (n.d.). The extinction crisis, https://bit.ly/3naKIYS.

Chamove, Arnold S. (1997). Dogs judge books by their covers. *Anthrozoös* 10(1): 50–52.

Channel Nine (2018). Sheep, ships and videotape: part one.' Channel Nine, *60 minutes*, https://bit.ly/32Nzl0N.

Cheke, Lucy G., and Nicola S. Clayton (2010). Mental time travel in animals. *WIREs cognitive science* 1: 915–930.

Chiesa, Alberto, Alessandro Serretti, and Janus Christian Jakobsen (2013). Mindfulness: top-down or bottom-up emotion regulation strategy?' *Clinical psychology review* 33: 82–96.

Clark, Corinna, Joanna Murrell, Mia Fernyhough, Treasa O'Rourke, and Michael Mendl (2014). Long-Term and trans-generational effects of neonatal experience on sheep behaviour. *Biology letters* 10(7), doi 10.1098/rsbl.2014.0273.

Clayton, Nicola S., and A. Dickson (2010). Mental time travel: can animals recall the past and plan for the future? In *Encyclopedia of animal behavior*. Michael D. Breed and Janice Moore, eds., 438–442. Elsevier.

Coan, James A. (2016). Toward a neuroscience of attachment. In *Handbook of attachment: theory, research, and clinical applications*. Jude Cassidy and Phillip R. Shaver, eds., 3rd edition, 242–269. New York: The Guilford Press.

Collins, Peter, and Anselma Gallinat (2010). *The ethnographic self as resource: writing memory and experience into ethnography*. Oxford: Berghahn Books.

Compa, Lance A. (2005). *Blood, sweat, and fear: workers' rights in U.S. meat and poultry plants*. New York: Human Rights Watch, https://digitalcommons.ilr.cornell.edu/articles/331/.

Cook, Peter F., Ashley Prichard, Mark Spivak, and Gregory S. Berns (2016). Awake canine fMRI predicts dogs' preference for praise vs food. *Social cognitive and affective neuroscience* 11(12): 1853–1862, https://doi.org/10.1093/scan/nsw102.

Cookson, Lawrence J. (2011). A definition for wildness. *Ecopsychology* 3(3): 187–193.

Corballis, Michael C. (2002). *From hand to mouth: the origins of language*. Princeton NJ: Princeton University Press.

Cordaro, Millie (2012). Pet loss and disenfranchised grief: implication for mental health counselling practice. *Journal of mental health counselling* 34(4): 283–294.

Corr, Charles A., Clyde Nabe, and Donna M. Corr (2006). *Death and dying, life and living*. 5th edition. Belmont, CA: Thomson Wadsworth.

Counted, Victor (2016). God as an attachment figure: a case study of the God attachment language and God concepts of anxiously attached Christian youths in South Africa.' *Journal of spirituality in mental health* 18(4): 316–346.

Cox, Marie, Eric Gaglione, Pamela Prowten, and Michael Noonan (1996). Food preferences communicated via symbol discrimination by a California sea lion (*Zalophus californianus*). *Aquatic mammals* 22(1): 3–10.

Craggs, Samantha (2016). Anita Krajnc pig trial: people are eating more and more meat, expert says. *CBC*, 10 November, https://bit.ly/3iyzOJd.

Craggs, Samantha (2017). Woman who gave water to pigs acted like Gandhi and Nelson Mandela: lawyer. *CBC*, 9 March, https://bit.ly/2GhMgQx.

Crandall, Floyd M. (1897). Hospitalism (Editorial). *Archives of paediatrics* 14(6): 448–454, https://bit.ly/2UvJLO4

Crook, Edgar (2014 [2008]). *Vegetarianism in Australia: a history.* Canberra: Self-published.

Crosby, Donald A. (2003). Transcendence and immanence in a religion of nature. *American journal of theology & philosophy* 24(3): 245–259.

Cyranoski, David (2010). Solitary fish hit rock bottom: 'frozen' zebrafish may be first piscene model for human depression. *Nature*, 22 November, doi:10.1038/news.2010.624.

D'Amico, Paola (2012). Il cetaceo che protegge il piccolo. *Corriere della sera*, 24 September, https://bit.ly/3cY2pGJ.

Dahlitz, Matthew (2014). An awareness of vicarious loss and grief in paramedic service. *Neuropsychotherapist*, 20 November, https://bit.ly/3niUMzj.

Damasio, Antonio (2006 [1994]). *Descartes' error.* London: Vintage Books.

Damasio, Antonio (1999). *The feeling of what happens: body and emotion in the making of consciousness.* New York: Harcourt Brace.

Damasio, Antonio (2010). *Self comes to mind: constructing the conscious brain.* London: Vintage Books.

Darwin, Charles (1872). *The expression of emotions in man and animals.* London: John Murray.

Dasgupta, Shreya (2015). Many animals can become mentally ill. *BBC Earth*, 9 September, https://bbc.in/3d1qw7k

Davidson, Richard J. (2004). Well-being and affective styles: neural substrates and biobehavioural correlates. *Philosophical transactions of the Royal Society* 359(1449): 1395–1411.

Davidson, Richard J., and Bruce S. McEwen (2012). Social influences on neuroplasticity: stress and interventions to promote wellbeing. *Nature neuroscience* 15(5): 689–695.

Davidson, Richard J., Jon Kabat-Zinn, Jessica Schumacher, Melissa Rozenkranz, Daniel Muller, Saki F. Santorelli, Ferris Urbanowski, Anne Harrington, Katherine Bonus, and John Sheridan (2003). Alterations in brain and immune function produced by mindfulness meditation. *Psychosomatic medicine* 65(4): 564–570.

Davis, Kingsley (1947). Final note on a case of extreme isolation. *American journal of sociology* 52(5): 432–437.

de Waal, Frans (1997). Are We in Anthropodenial? *Discovery magazine*, 1 July, https://bit.ly/3f9olj5.

de Waal, Frans (2013). *The bonobo and the atheist: in search of humanism among the primates*. New York: W.W. Norton & Co.

de Waal, Frans (2016). *Are we smart enough to know how smart animals are?* London: Granta.

de Waal, Frans (2019). *Mama's last hug: animal emotions and what they teach us about ourselves*. London: Granta.

DeGregorio, Lisa J. (2013). Intergenerational transmission of abuse: implications for parenting interventions from a neuropsychological perspective. *Traumatology* 19(2): 158–166.

Del Rio, Carlos M., and Lyle J. White (2012). Separating spirituality from religiosity: a hylomorphic attitudinal perspective. *Psychology of religion and spirituality* 4(2): 123–142.

Derbyshire, David (2009). Magpies grieve for their dead (and even turn up for funerals). *Daily Mail*, 24 October, http://dailym.ai/33sQuxo.

Despret, Vinciane (2009). Culture and gender do not dissolve into how scientists "read" nature: Thelma Rowell's heterodoxy. In *Rebels, mavericks, and heretics in biology*. Oren Harman, ed., 338–355. New Haven: Yale University Press.

Despret, Vinciane (2005). Sheep do have opinions. In *Making things public: atmospheres of democracy*. Bruno Latour and Peter Weibel, eds., 360–370. Cambridge: M.I.T. Press.

Destro, Anna Maria (2009). Grief, bereavement, and mourning in historical perspective. *Encyclopedia of death and the human experience*. Clifton D. Bryant and Dennis L. Peck, eds., 533–538. Thousands Oaks, CA: Sage Publications.

Dillard, Jennifer (2008). Slaughterhouse nightmare: psychological harm suffered by slaughterhouse employees and the possibility of redress through legal reform. *Georgetown journal on poverty law and policy* 15(2): 391–408.

Donaldson, Sue, and Will Kymlicka (2015). Farmed animal sanctuaries: the heart of the movement? *Politics and animals* 1(1): 50–74.

Donoghue, John D. (1957). An Eta community in Japan: the social persistence of outcaste groups. *American anthropologist* 59(6): 1000–1017.

Dosa, David M. (2007). A day in the life of Oscar the cat. *The New England journal of medicine* 357: 328–329, doi: 10.1056/NEJMp078108.

Droseltis, Orestis, and Vivian L. Vignoles (2010). Towards an integrative model of place identification: dimensionality and predictors of intrapersonal-level place preferences. *Journal of environmental psychology* 30(1): 23–34.

Works cited

Dubuisson, Daniel (2003). *The Western construct of religion*. Translated from the French original (1998) by William Sayers. Baltimore: The Johns Hopkins University Press.

Dugdale Henrietta (1883). *A few hours in a far-off age*. Melbourne: M'Carron, Birds & Co.

Dwyer Cathy (2009). The behaviour of sheep and goats. In *The ethology of domestic animals: an introductory text*. Per Jensen, ed., 2nd edition, 161–174. Wallingford (UK): CABI.

Ehrlich, Katherine B., Gregory E. Miller, Jason D. Jones, and Jude Cassidy (2016). Attachment and psychoneuroimmunology. In *Handbook of attachment: theory, research, and clinical applications*. Jude Cassidy and Phillip R. Shaver, eds., 3rd edition, 180–201. New York: The Guilford Press.

Eisenberger, Naomi I. (2015). Meta-analytic evidence for the role of the anterior cingulate cortex in social pain. *Social cognitive and affective neuroscience* 10(1): 1–2, doi: 10.1093/scan/nsu120.

Engler, Steven, and Dean Miller (2006). Daniel Dubuisson, The western construction of religion. *Religion* 36: 119–178.

Englund, Harri (1998). Death, trauma and ritual: Mozambican refugees in Malawi.' *Social science and medicine* 46(9): 1165–1174.

Erlandson, Sven (2000). *Spiritual but not religious: a call to religious revolution in America*. Bloomington: iUniverse.

Erlangsen, Annette, Bernard Jeune, Unni Bille-Hrahe, and James W. Vaupel (2004). Loss of a partner and suicide risks among oldest old: a population-based register study. *Age and ageing* 33(4): 378–383.

Farb, Norman A.S., Zindel V. Segal, Helen Mayberg, Jim Bean, Deborah McKeon, Zainab Fatima, and Adam K. Anderson (2007). Attending to the present: mindfulness meditation reveals distinct neural modes of self-reference. *Social cognitive & affective neuroscience* 2(4): 313–322.

Faublée, Jacques (1947). *Récits bara*. Paris: Institut d'Ethologie.

Feldman Barrett, Lisa, Kristen A. Lindquist, Eliza Bliss-Moreau, Seth Duncan, Maria Gendron, Jennifer Mize, and Lauren Brennan (2007). Of mice and men: natural kinds of emotions in the mammalian brain? A response to Panksepp and Izard. *Perspectives on psychological science* 2(3): 297–311.

Felman, Shoshana, and Dori Laub (1992). *Testimony: crisis of witnessing in literature, psychoanalysis, and history*. New York: Routledge.

Ferdowsian, Hope (2018). *Phoenix zones: where strength is born and resilience lives*. Chicago: The University of Chicago Press.

Ferdowsian, Hope, and Debra Merskin (2012). Parallels in sources of trauma, pain, distress, and suffering in humans and nonhuman animals. *Journal of trauma & dissociation* 13(4): 448–468.

Field, Tiffany (2011). Romantic breakups, heartbreak and bereavement. *Psychology* 2(4): 382–387.

Fisher, Christopher L. (2005). Animals, humans and x-men: human uniqueness and the meaning of personhood. *Theology and science* 3(3): 291–314.

Fitzgerald, Amy J. (2010). A social history of the slaughterhouse: from inception to contemporary implications. *Human ecology review* 17(1): 58–69.

Fitzgerald, Amy J., Linda Kalof, and Thomas Dietz (2009). Spillover from 'The Jungle' into the larger community: slaughterhouses and increased crime rates. *Organization and environment* 22(2): 158–184.

Flanagan, Cara (1999). *Early socialisation: sociability and attachment.* London, New York: Routledge.

Formosa, Amy (2016). Photographer captures kangaroo family's grief. *Fraser Coast Chronicle*, 13 January, https://bit.ly/36XvLdD.

Fraley, Chris R., and Phillip R. Shaver (2016). Attachment, loss, and grief: Bowlby's views, new developments, and current controversies. *Handbook of attachment: theory, research, and clinical applications.* Jude Cassidy and Phillip R. Shaver, eds., 3rd edition, 40–62. New York: The Guilford Press.

Fraley, R. Chris, Nathan W. Hudson, Marie E. Heffernan, and Noam Segal (2015). Are adult attachment styles categorical or dimensional? A taxometric analysis of general and relationship-specific attachment orientations. *Journal of personality and social psychology* 109(2): 354–368.

Frazer, James George (1922). *The golden bough.* New York: Macmillan, http://www.bartleby.com/196/.

Freud, Anna, and Sophie Dann (1951). An experiment in group upbringing. *Psychoanalytic study of the child* 6: 127–168.

Gainotti, Guido (2012). Unconscious processing of emotions and the right hemisphere. *Neuropsychologia* 50: 205–218.

Gallate, Jason, Cara Wong, Sophie Ellwood, Richard Chi, and Allan Snyder (2011). Noninvasive brain stimulation reduces prejudice scores on an implicit association test. *Neuropsychology* 25(2): 185–192.

Gander, Manuela, and Anna Buchheim. 'Attachment classification, psychophysiology and frontal EEG asymmetry across the lifespan: a review.' *Frontiers in human neuroscience* 9(79), doi: 10.3389/fnhum.2015.00079.

Garber, Megan (2013). Animal behaviorist: we'll soon have devices that let us talk with our pets. *The Atlantic*, 4 June, https://bit.ly/3nhIbwj.

Garland, Eric L., Brett Froelinger, and Matthew O. Howard (2015). Neurophysiological evidence for remediation of reward processing deficits in chronic pain and opioid misuse following treatment with mindfulness-oriented recovery enhancement: exploratory ERP findings from a pilot RCT. *Journal of behavioral medicine* 38(2): 327–336.

Works cited

Gaynor, Kaitlyn M., Cheryl E. Hojnowski, Neil H. Carter, and Justin S. Brashares
(2018). The influence of human disturbance on wildlife nocturnality. *Science*
360(6394): 1232–1235.
Gaynor, Kaitlyn (2018). To avoid humans, more wildlife now work the night shift.
The conversation, 15 June, https://bit.ly/3nlRAmn.
Gazzola, Lauren (2014). Animals die, woman cries, but we are unmoved. *Truthout*,
23 October, https://bit.ly/33y106k.
Gazzola, Valeria, Giacomo Rizzolatti, Bruno Wicker, and Christian Keysers (2007).
The anthropomorphic brain: the mirror neuron system responds to human
and robotic actions. *NeuroImage* 35(4): 1674–1684.
Geertz, Clifford (1975). On the nature of anthropological understanding.
American scientist 63(1): 47–53.
Gonçalves, André, and Dora Biro (2018). Comparative thanatology, an integrative
approach: exploring sensory/cognitive aspects of death recognition in
vertebrates and invertebrates. *Philosophical transactions of the Royal Society B*
373(1754): 20170263.
Gonzalez Fischer, Carlos, and Tara Garnett (2016). *Plates, pyramids and planets –
developments in national healthy and sustainable dietary guidelines: a state of
play assessment*. FAO and University of Oxford, http://www.fao.org/3/
a-i5640e.pdf.
Goodall, Jane (2006). The dance of awe. *A communion of subjects: animals in
religion, science & ethics*. Paul Waldau and Kimberley Patton, eds., 651–656.
New York: Columbia University Press.
Goodson, James L., Aubrey M. Kelly, and Marcy A. Kingsbury (2012). Evolving
nonapeptide mechanisms of gregariousness and social diversity in birds.
Hormones and behavior 61(3): 239–250.
Graham, Linda (2008). The neuroscience of attachment. Paper presented as a
Clinical Conversation at the Community Institute for Psychotherapy. San
Rafael, CA, https://bit.ly/36I7S1H.
Graham, Lindsay T., Samuel D. Gosling, and Christopher K. Travis (2015). The
psychology of home environments: a call for research on residential space.
Perspectives on psychological science 10(3): 346–356.
Grandin, Temple, and Catherine Johnson (2005). *Animals in translation: using the
mysteries of autism to decode animal behavior*. New York: Scribner.
Granqvist, Pehr, Mario Mikulincer, Vered Gewirtz, and Phillip R. Shaver (2012).
Experimental findings on God as attachment figure: normative processes and
moderating effects of internal working models. *Journal of personality and
social psychology* 103(5): 804–818.
Granqvist, Pehr (2002). Attachment and religion: an integrative developmental
framework. Ph.D. thesis, Uppsala University, Sweden, https://bit.ly/36zhnS8.

Greenway, Robert (2009). The *Ecopsychology* interview. *Ecopsychology* 1(1): 47–52.

Grieser, Alexandra, and Jay Johnston, eds., (2017). *Aesthetics of religion: a connective concept.* Berlin: De Gruyter.

Griffin, Donald R. (2006). From cognition to consciousness. *A communion of subjects: animals in religion, science and ethics.* Paul Waldau and Kimberley Patton, eds., 481–504. New York: Columbia University Press.

Griffin, Donald R. (2001 [1992]). *Animal minds: beyond cognition to consciousness.* Chicago: University of Chicago Press.

Griffiths, Scott K., W.S. Brown Jr., Kenneth J. Gerhardt, Robert M. Adams, and Richard J. Morris (1994). The perception of speech sounds recorded within the uterus of a pregnant sheep. *Journal of the Acoustical Society of America* 96(4): 2055–2063.

Gross, Aaron S. (2015). *The question of the animal and religion.* New York: Columbia University Press.

Grüsser, Otto-Joachim, Thomas Selke, and Barbara Zynda (1988). Cerebral lateralization and some implications for art, aesthetic perception, and artistic creativity. In *Beauty and the brain: biological aspects of aesthetics.* Ingo Rentschler, Barbara Herzberger and David Epstein, eds., 257–293. Basel: Birkhauser Verlag.

Grüsser, Otto-Joachim (1983). Mother-child holding patterns in Western art: a developmental study. *Ethology and sociobiology* 4(2): 89–94.

Gündel, H., M.F. O'Connor, L. Littrell, C. Fort, and R.D. Lane (2003). Functional neuroanatomy of grief: an fMRI study. *The American journal of psychiatry* 160(1): 1946–1953.

Grušovnik, Tomaž (2018). Debordering ethics: acknowledging animal morality. In *Borders and debordering: topologies, praxes, hospitableness.* Tomaž Grušovnik, Eduardo Mendieta and Lenart Škof, eds., 133–148 . London: Lexington Books.

Güntürkün, Onur (2005). The avian 'prefrontal cortex' and cognition. *Current opinion in neurobiology* 15(6): 686–693.

Gustafsson, Bengt (1997). The health and safety of workers in a confined animal system. *Livestock production science* 49(2): 191–202.

Guthrie, Stewart E. (1980). A cognitive theory of religion. *Current anthropology* 21(2): 181–203.

Guthrie, Stewart (2002). Animal animism: evolutionary roots of religious cognition. In *Current approaches in the cognitive science of religion.* Ilkka Pyysiäinen and Veikko Anttonen, eds., 38–67. London, New York: Continuum.

Guthrie, Stewart (1993). *Faces in the clouds: a new theory of religion.* New York, Oxford: Oxford University Press.

Gutmann Joseph (1998). On Biblical legends in Medieval art. *Artibus et Historiae* 19(38): 137–142.

Hamilton, C. R., and B.A. Vermeire (1988). Complementary hemispheric specialization in monkeys. *Science* 242(4886): 1691–1694.

Han, Shihui, and Georg Northoff (2009). Understanding the self: a cultural neuroscience approach. *Progress in brain research.* J.Y. Chiao, ed., vol. 178, Chapter 14, pp. 203–212.

Hardus, Madeleine E., Adriano R. Lameira, Carel P. Van Schaik, and Serge A. Wich (2009). Tool use in wild orang-utans modifies sound production: a functionally deceptive innovation? *Proceedings of the Royal Society B*, doi: 10.1098/rspb.2009.1027.

Harlow, Harry F. (1958). The nature of love. *American psychologist* 13: 673–685.

Harlow, Harry F., and Margaret Kuenne Harlow (1962). Social deprivation in monkeys. *Scientific American* 5: 136–146.

Harlow, Harry F., Robert O. Dodsworth, and Margaret K. Harlow (1965). Total social isolation in monkeys. *Proceedings of the National Academy of Sciences USA* 54(1): 90–97.

Harris, L.J., J.B. Almerigi, T.J. Carbary, and T.G. Fogel (2001). Left-side infant holding: a test of the hemispheric arousal-attentional hypothesis. *Brain and cognition* 46(1-2): 159–165.

Harrod, James B. (2011). A trans-species definition of religion. *Journal for the study of religion, nature and culture* 5(3): 327–353.

Harrod, James B. (2014). The case for chimpanzee religion. *Journal for the study of religion, nature and culture* 8(1): 8–45.

Harrod, James B. (2016). Conceiving animacy: a group-theoretic transformation formula defining spirituality, based on neuroscience and attuned to global etymologies. Presented at *Religion, science and the future,* The International Society for the study of religion, nature and culture, The University of Florida, 14–17 January.

Harvey, Graham (2005). *Animism: respecting the living world.* London: Hurst & Company.

Hatkoff, Amy (2009). *The inner world of farm animals.* New York: Stewart, Tabori & Chang.

Hazan, Cindy, and Phillip Shaver (1987). Romantic love conceptualized as an attachment process. *Journal of personality and social psychology* 52(3): 511–524.

Henrich, Joseph, Steven J. Heine, and Ara Norenzayan (2010). The weirdest people in the world? *Behavioural and brain sciences* 33(2-3): 61–83.

Heuts, Boudewijn Adriaan, and Tibor Brunt (2005). Behavioral left-right asymmetry extends to arthropods. *Behavioral and brain sciences* 28(4): 601–602.

Higginbotham, Nick, Linda Connor, Glen Albrecht, Sonia Freeman, and Kingsley Agho (2006). Validation of an environmental distress scale. *EcoHealth* 3(4): 245–254.

Hilger, M. Inez (1951). *Chippewa child life and its cultural background.* Washington, D.C.: US Government Print Office.

Hill, Peter C., Kenneth I. Pargament, Ralph W. Hood Jr, Michael E. McCullough, James P. Swyers, David B. Larson, and Brian J. Zinnbauer (2000). Conceptualising religion and spirituality: points of commonality, points of departure. *Journal for the theory of social behaviour* 20(1): 51–77.

Hillman, James (1996). Aesthetics and politics. *Tikkun* 11(6): 38–40.

Hinds, Joe, and Paul Sparks (2008). Engaging with the natural environment: the role of affective connection and identity. *Journal of environmental psychology* 28(2): 109–120.

Hirshon, Bob (n.d.). Not dead yet. *Science net links*, https://bit.ly/3iBllfs.

Hockey, Jenny, Jeanne Katz, and Neil Small, eds., (2001). *Grief, mourning and death ritual.* Philadelphia: Open University Press.

Hofer, Myron A. (1984). Relationships as regulators: a psychobiological perspective on bereavement. *Psychosomatic medicine* 46(3): 183–197.

Hofer, Myron A. (2006). Psychobiological roots of early attachment. *Current directions in psychological science* 15(2): 84–88.

Holland, Jennifer (2011). *Unlikely friendships: 47 remarkable stories from the animal kingdom.* New York: Workman.

Honeyborne, James (2013). Elephants really do grieve like us: they shed tears and even try to 'bury' their dead – a leading wildlife film-maker reveals how the animals are like us. *Daily Mail online*, 31 January, http://dailym.ai/3leTuDI.

Hooper, Rowan (2011). Death in dolphins: do they understand they are mortal? *New scientist online*, 1 September, https://bit.ly/2Gu3ydd.

Horner, John R., and Robert Makela (1979). Nest of juveniles provides evidence of family structure among dinosaurs. *Nature* 282(5736): 296–298.

Horowitz, Alexandra C., and Marc Bekoff (2007). Naturalizing anthropomorphism: behavioural prompts to our humanizing of animals. *Anthrozoös* 20(1): 23–35.

Huntington, Richard, and Peter Metcalf (1979). *Celebration of death: the anthropology of mortuary ritual.* Cambridge: Cambridge University Press.

Inbar, Michael (2010). Mom's hug revives baby that was pronounced dead. *Today*, 4 September, https://on.today.com/2HPPqv0

Infoplease (n.d.). Gestation, incubation, and longevity of selected animals. *Infoplease.* © 2000-2017 Sandbox Networks, Inc., publishing as Infoplease, https://bit.ly/3JxnaLN

Ingold, Tim (2000). *The perception of the environment.* London: Routledge.

Innis Dagg, Anne (2011). *Animal friendships.* Cambridge: Cambridge University Press.

Insel, Thomas R. (2010). The challenge of translation in social neuroscience: a review of oxytocin, vasopressin, and affiliative behaviour. *Neuron* 65(6): 768–779, doi: https://doi.org/10.1016/j.neuron.2010.03.005.

Ironside, Virginia (2011). Romania's orphanages: locking the past away. *The Independent*, 29 November, https://bit.ly/33ui6SE.

Johnson, Mark (2007). *The meaning of the body: aesthetics of human understanding.* Chicago: The University of Chicago Press.

Johnston, Jay (2017). Rewilding religion: affect and animal dance. *Bulletin for the study of religion* 46(3-4): 11–16.

Johnstone, Brick, Angela Bodling, Dan Cohen, Shawn E. Christ, and Andrew Wegrzyn (2012). Right parietal lobe-related 'selflessness' as the neuropsychological basis of spiritual transcendence. *International journal for the psychology of religion* 22(4): 267–284.

jones, pattrice (2007). *Aftershock: confronting trauma in a violent world, a guide for activists and their allies.* New York: Lantern Books.

jones, pattrice (2010). Harbingers of (silent) spring: archetypal avians, avian archetypes and the truly collective unconscious. *Spring* 83: 185–212.

jones, pattrice (2017). What I've learned from animals at VINE sanctuary. Presentation at the Farm Sanctuary Hoe Down, Watkins Glen (NY), 12–13.

Jones, Robert (1888). *Vegetarianism, with special reference to its connection with temperance in drinking.* Lecture given at the Total Abstinence Society, Melbourne, 10 April.

Jonker, Gerdien (1997). The many facets of Islam: death, dying and disposal between orthodox rule and historical convention. In *Death and bereavement across cultures.* Colin Murray Parkes, Pittu Laungani and Bill Young, eds., 147–165. London: Routledge.

Josephson, Brian D., and Fotini Pallikari-Viras (1991). Biological utilisation of quantum nonlocality. *Foundations of physics* 21: 197–207, https://bit.ly/35yRsJz.

Jost, John T., and Mahzarin R. Banaji (1994). The role of stereotyping in system-justification and the production of false consciousness. *British journal of social psychology* 33(1): 1–27.

Jost, John T., Mahzarin R. Banaji, and Brian A. Nosek (2004). A decade of system justification theory: accumulated evidence of conscious and unconscious bolstering of the status quo. *Political psychology* 25(6): 881–919.

Joy, Melanie (2009). *Why we love dogs, eat pigs and wear cows*. San Francisco: Conari Press.

Juhl, Jacob, and Clay Routledge (2014). Finding the terror that the social self manages: interdependent self-construal protects against the anxiety engendered by death awareness. *Journal of social and clinical psychology* 33(4): 365–379.

Kahn, Peter H. Jr. (2009). Cohabiting with the wild. *Ecopsychology* 1(1): 38–46.

Kaprio, J., M. Koskenvuo, and H. Rita (1987). Mortality after bereavement: a prospective study of 95,647 widowed persons. *American journal of public health* 77(3): 283–287.

Karenina, Karina, Andrey Giljov, Vladimir Baranov, Ludmila Osipova, Vera Krasnova, and Yegor Malaschiev (2010). Visual laterality of calf-mother interactions in wild whales? *PLOS ONE* 5(11): e13787.

Kastenbaum, Robert (1987). Vicarious grief: an intergenerational phenomenon? *Death studies* 11(6): 447–453.

Kaufman, Leslie (2012). Date night at the zoo, if rare species play along. *The New York Times*, 4 July, https://nyti.ms/3nj7vBZ.

Kellehear, Allan (2008). Dying as a social relationship: a sociological review of debates on the determination of death. *Social science and medicine* 66(7): 1533–1544.

Keller, Helen (1929). *We bereaved*. New York: Leslie Fulenwider, Inc. Publishers, https://bit.ly/32IM0C3.

Kelley, Carl F. (2009 [1997]). *Meister Eckhart on divine knowledge*. New Haven: Yale University Press.

Keysers, Christian, and Valeria Gazzola (2010). Social neuroscience: mirror neurons recorded in humans. *Current biology* 20(8): R353–354.

Kierkegaard, Søren (1980 [1841]). *The concept of anxiety: a simple psychologically orienting deliberation on the dogmatic issue of hereditary sin*. Edited and translated by Reidar Thomte in collaboration with A.B. Anderson. Princeton: Princeton University Press.

King, Barbara (2013). *How animals grieve*. Chicago: The University of Chicago Press.

King, Stephanie L., and Vincent M. Janik (2013). Bottlenose dolphins can use learned vocal labels to address each other. *Proceedings of the National Academy of Sciences USA* 110(32): 13216–13221.

Kirmayer, Laurence J., Christopher Fletcher, and Robert Watt (2008). Locating the ecocentric self: Inuit concepts of mental health and illness. In *Healing traditions: the mental health of Aboriginal peoples in Canada.* Laurence J. Kirmayer and Gail Guthrie Valaskakis, eds., Vancouver: University of British Columbia Press.

Knobloch, H.S., A. Charlet, L.C. Hoffman, M. Eliava, S. Khrulev, A.H. Cetin, P. Osten, M.K. Schwarz, P.H. Seeburg, R. Stoop, and V. Grinevich (2012). Evoked axonal oxytocin release in the central amygdala attenuates fear response. *Neuron* 73(3): 553–566, doi: 10.1016/j.neuron.2011.11.030.

Knutson, B., C.M. Adams, G.W. Fong, and D. Hommer (2001). Anticipation of increasing monetary reward selectively recruits nucleus accumbens. *Journal of neuroscience* 21(16): RC159.

Koga, Kazuko, and Yutaka Iwasaki (2013). Psychological and physiological effect in humans of touching plant foliage – using the semantic differential method and cerebral activity as indicators. *Journal of physiological anthropology* 32(1): 7, doi: 10.1186/1880-6805-32-7.

Koluchova, Jarmila (1972). Severe deprivation in twins: a case study. *Journal of child psychology and psychiatry* 13: 107–114.

Korpela, Kalevi Mikael (1989). Place-identity as a product of environmental self-regulation. *Journal of environmental psychology* 9(3): 241–256.

Krämer, Hans Martin (2008). 'Not benefitting our divine country': eating meat in Japanese discourses of self and other from the seventeenth century to the present.' *Food and foodways* 16(1): 33–62.

Kramer, Matthew, and Gordon M. Burghardt (1998). Precocious courtship and play in emydid turtles. *Ethology* 104: 38–56.

Kristeva, Julia (1982). *Powers of horror: an essay on abjection.* Translated by Leon S. Roudiez. New York: Columbia University Press.

Kubinyi, Eniko, Ádám Miklósi, Frédéric Kaplan, Márta Gácsi, József Topál, and Vilmos Csányi (2004). Social behaviour of dogs encountering AIBO, an animal-like robot in a neutral and in a feeding situation. *Behavioural processes* 65(3): 231–239.

Lancy, David F. (2014). 'Babies aren't persons': a survey of delayed personhood. In *Different faces of attachment: cultural variations of a universal human need.* Heidi Keller and Otto Hiltrud, eds., 66–109. Cambridge: Cambridge University Press.

Latour, Bruno (2010). *On the cult of the factish gods.* Durham and London: Duke University Press.

LeDoux, Joseph (2012). Rethinking the emotional brain. *Neuron* 73(4): 653–676.

Lewicka, Maria (2011). Place attachment: how far have we come in the last 40 years? *Journal of environmental psychology* 31(3): 207–230, https://doi.org/10.1016/j.jenvp.2010.10.001.

Lifton, Robert Jay (1986). *The Nazi doctors: medical killing and the psychology of genocide*. New York: Basic Books.

Linzey, Andrew, and Dan Cohn-Sherbok (1997). *After Noah: animals and the liberation of theology*. Herndon, VA: Mowbray.

Liu, Dong, Josie Diorio, Jamie C. Day, Darlene D. Francis, and Michael J. Meaney (2000). Maternal care, hippocampal synaptogenesis and cognitive development in rats. *Nature neuroscience* 3(8): 799–806.

Lockwood, Alex (2018). Bodily encounter, bearing witness and the engaged activism of the global Save movement. *Animal studies journal* 7(1): 104–126.

Longbottom, Sarah, and Virginia Slaughter. Sources of children's knowledge about death and dying. *Philosophical transactions of the Royal Society B* 373(1754): 20170267.

Lorenz, Konrad (1937). The companion in the bird's world. *The auk* 54: 245–273, abbreviated version of the original paper published in German in 1935: 'Der Kumpan in der Umwelt des Vogels.' *Journal fur ornithologie* 83: 137–213.

Ludden, Jennifer (2005). A hippo and tortoise tale. *NPR*, 17 July, https://www.npr.org/templates/story/story.php?storyId=4754996.

MacDonald, Geoff, and Mark R. Leary (2005). Why does social exclusion hurt? The relationship between social and physical pain. *Psychological bulletin* 131(2): 202–223.

MacNair, Rachel M. (2002). *Perpetration-induced traumatic stress: the psychological consequences of killing*. Praeger/Greenwood Publishing Group.

Maestripieri, Dario (2005). Early experience affects the intergenerational transmission of infant abuse in rhesus monkeys. *Proceedings of the National Academy of Sciences* 102: 9726–9729.

Maier, Steven F., and Martin E.P. Seligman (2017). Learned helplessness at fifty: insights from neuroscience. *Psychological review* 123(4): 349–367.

Main, Mary, and Judith Solomon (1986). Discovery of an insecure-disorganised/disoriented attachment pattern. In *Affective development in infancy*. T. Berry Brazelton and Michael W. Yogman, eds., 95–124. Norwood, NJ: Ablex Publishing Co.

Mallon, Brenda (2008). *Dying, death and grief: working with adult bereavement*. Los Angeles: Sage Publications.

Mancia, Mauro (2006). Implicit memory and early unrepressed unconscious: their role in the therapeutic process (how the neurosciences can contribute to psychoanalysis). *The international journal of psychoanalysis* 87: 83–103.

Manning, J.T., R. Heaton, and A.T. Chamberlain (1994). Left-side cradling: similarities and differences between apes and humans. *Journal of human evolution* 26(1): 77–83.

Mapes, Lynda V. (2018). After 17 days and 1,000 miles, mother orca Tahlequah drops her dead calf. *The Seattle Times*, 11 August, https://bit.ly/3ni8fHr.

Marder, Michael (2013a). Is it ethical to eat plants? *Parallax* 19(1): 29–37.

Marder, Michael (2013b). Should plants have rights? *The philosophers' magazine* 61: 46–50.

Marino, Lori, and Christina M. Colvin (2015). Thinking pigs: a comparative review of cognition, emotion, and personality in *Sus domesticus*. *International journal of comparative psychology* 28(1), article ID 23859.

Marino, Lori (2017). Thinking chickens: a review of cognition, emotion, and behaviour in the domestic chicken. *Animal cognition* 20(2): 127–147.

Marino, Lori, and Kristin Allen (2017). The psychology of cows. *Animal behavior and cognition* 4(4): 474–498.

Marino, Lori, and Debra Merskin (2019a). Intelligence, complexity, and individuality in sheep. *Animal sentience* 2019.206.

Marino, Lori, and Debra Merskin (2019b). Deepening our understanding of sheep. *Animal sentience* 2019.276.

Marino, Lori, and Michael Mountain (2015). Denial of death and the relationship between humans and other animals. *Anthrozoös* 28(1): 5–21.

Mark, Patty (2014). Dreams and beyond. Interviewed by TBP. *Southerly* 74(3): 102–116.

Mark, Patty (n.d.). No title, https://bit.ly/32PP4fw.

Markus, Hazel Rose, and Shinobu Kitayama (1991). Culture and the self: implications for cognition, emotion, and motivation. *Psychological review* 98(2): 224–253.

Martin, Dan (1996). On the cultural ecology of sky burial on the Himalayan plateau. *East and west* 46(3-4): 353–370.

Masson, Jeffrey M. (2009). *The face on your plate*. New York: W.W. Norton & Co, Inc.

Masson, Jeffrey M. (2020). *Lost companions: reflections on the death of pets*. New York: St. Martin's Press.

Mathews, Freya (2013). Against kangaroo harvesting. *Bioethical inquiry* 10(2): 263–265.

Mathis, Clay P., and Boone Carter (2008). Minimizing weaning stress on calves. *Guide B-221*. Cooperative extension service, College of Agriculture and Home Economics, New Mexico State University, https://bit.ly/33tIhJl.

McCann, I. Lisa, and Laurie Anne Pearlman (1990). Vicarious traumatization: a framework for understanding the psychological effects of working with victims. *Journal of traumatic stress* 3(1): 131–149.

McGilchrist, Iain (2009). *The master and his emissary: the divided brain and the making of the western world*. New Haven: Yale University Press.

McLaughlin, Katie A., Margaret A. Sheridan, Sonia Alves, and Wendy Berry Mendes (2014). Child maltreatment and autonomic nervous system reactivity: identifying dysregulated stress reactivity patterns using the biopsychosocial model of challenge and threat. *Psychosomatic medicine* 76(7), doi: 10.1097/PSY.0000000000000098.

Meaney, Michael J. (2001). Maternal care, gene expression, and the transmission of individual differences in stress reactivity across generations. *Annual review of neuroscience* 24: 1161–1192.

Meijer, Eva (2019 [2016]). *Animal languages: the secret conversations of the living world.* Translated by Laura Watkinson. London: John Murray.

Mellor, D.J. (2015). Positive animal welfare states and encouraging environment-focused and animal-to-animal interactive behaviours. *New Zealand veterinary journal* 63(1): 9–16.

Mendl, Michael, Elizabeth S. Paul, and Lars Chittka (2011). Animal behaviour: emotion in invertebrates? *Current biology* 21(12): R463–465.

Merker, Bjorn (2007). Consciousness without a cerebral cortex: a challenge for neuroscience. *Behavioural and brain sciences* 30(1): 63–81; discussion 81–134.

Merleau-Ponty, Maurice (2005 [1958, 1945]). *Phenomenology of perception.* Translated by Colin Smith. London & New York: Routledge.

Merskin, Debra (2004). The construction of Arabs as enemies: Post September 11 discourse of George W. Bush. *Mass communication & society, 7*(2), 157–175.

Merskin, Debra (2018). *Seeing species: re-presentations of animals in media & popular culture.* New York: Peter Lang.

Midgley, Mary (2002 [1978]). *Beast and man: the roots of human nature.* London: Routledge.

Mikulincer, Mario, and Phillip R. Shaver (2016). Adult attachment and emotion regulation. In *Handbook of attachment: theory, research, and clinical applications.* Jude Cassidy and Phillip R. Shaver, eds., 3rd edition, 507–533. New York: The Guilford Press

Mitrovic, Igor (n.d.). *Introduction to the hypothalamo-pituitary-adrenal (HPA) axis.* San Francisco: Biochemistry and Biophysics, UCSF, https://bit.ly/3lcL67B.

Mjaaland, Thera (2004). Ane Sugh' Ile. I keep quiet. Master's thesis, University of Bergen, Bergen, Norway.

Moon, Christine, Hugo Lagercrantz, and Patricia K. Kuhl (2013). Language experience *in utero* affects vowel perception after birth: a two-country study. *Acta paediatrica* 102(2): 156–160.

Moore, Ernest O. (1981). A prison environment's effect on health care service demands. *Journal of environmental systems* 11(1): 17–34.

Moorman, Sanne, and Alister U. Nicol (2015). Memory-related brain lateralisation in birds and humans. *Neuroscience and biobehavioral reviews* 50: 86–102.

Morgan, Paul (2010). Towards a developmental theory of place attachment. *Journal of environmental psychology* 30(1): 11–22.

Morris, Richard, and Marianne Fillenz (2003). *Neuroscience: the science of the brain*. Liverpool: The British Neuroscience Association.

Morse, Melvin L. (1994). Near death experience and death-related visions in children: implications for the clinician. *Current problems in paediatrics* 24: 55–83.

Mowat, Philippa (2009). Meat and modernity: changing perceptions of beef in the making of modern Japan. *Cross-sections* 5: 41–55.

National Research Council US (2009). *Recognition and alleviation of pain in laboratory animals*. Washington, D.C.: The National Academies Press, https://bit.ly/36E7kK3.

Negoias, S., I. Croy, J. Gerber, S. Puschmann, K. Petrowski, P. Joraschky, and T. Hummel (2010). Reduced olfactory bulb volume and olfactory sensitivity in patients with acute major depression. *Neuroscience* 169(1): 415–421, https://doi.org/10.1016/j.neuroscience.2010.05.012.

Nelson, Eric E., and Jaak Panksepp (1998). Brain substrates of infant-mother attachment: contributions of opioids, oxytocin, and norepinephrine. *Neuroscience and biobehavioral reviews* 22(3): 437–452.

Nelson, Kevin (2011). *The spiritual doorway in the brain: a neurologist's search for the god experience*. New York: Dutton.

Nett, Randall J., Tracy K. Witte, Stacy M. Holzbauer, Brigid L. Elchos, Enzo R. Campagnolo, Karl J. Musgrave, Kris K. Carter, Katie M. Kurkjian, Cole Vanicek, Daniel R. O'Leary, Kerry R. Pride, and Renee H. Funk (2015). Notes from the field: prevalence of risk factors for suicide among veterinarians – United States, 2014. *Morbidity and mortality weekly report* 64(5): 131–132, https://bit.ly/3kyFW52.

Newkey-Burden, Chas (2017). Our counter-terrorism experts shouldn't be wasting their time on the 'Save' animal welfare protest movement. *The Independent*, 14 February, https://bit.ly/34pcc4B.

Nolte, D.L., F.D. Provenza, and D.F. Balph (1990). The establishment and persistence of food preferences in lambs exposed to selected foods. *Journal of animal science* 68(4): 998–1002.

Nordanger, Dag Ø. (2007). Coping with loss and bereavement in post-war Tigray, Ethiopia. *Transcultural psychiatry* 44(4): 545–565.

Northoff, Georg, and Jaak Panksepp (2008). The trans-species concept of self and the subcortical-cortical midline system. *Trends in cognitive sciences* 12(7): 259–264.

Northoff, Georg, Alexander Heinzel, Moritz de Greck, Felix Bermpohl, Henrik Dobrowoly, and Jaak Panksepp (2006). Self-referential processing in our brain – a meta-analysis of imaging studies on the self. *NeuroImage* 31(1): 440–457.

Norton, Michael I., and Francesca Gino (2014). Rituals alleviate grieving for loved ones, lovers, and lotteries. *Journal of experimental psychology: general* 143(1): 266–272.

Nowbahari, Elise, Alexandra Scohier, Jean-Luc Duran, and Karen L. Hollis (2009). Ants, *cataglyphis cursor*, use precisely directed rescue behavior to free entrapped relatives. *PLOS ONE* 4(8): e6573, https://doi.org/10.1371/journal.pone.0006573.

O'Connor, Mary-Frances, David K. Wellisch, Annette L. Stanton, Naomi I. Eisenberger, Michael R. Irwin, and Matthew D. Lieberman (2008). Craving love? Enduring grief activates brain's reward center. *NeuroImage* 42(2): 969–972.

O'Connor, Mary-Frances (2012). Immunological and neuroimaging biomarkers of complicated grief. *Dialogues in clinical neuroscience* 14(2): 141–148.

Oliver, Kelly (2011). Pet lovers, pathologized. *New York Times*, 30 October, https://nyti.ms/3d1vIYS.

Oliver, Kelly (2001). *Witnessing: beyond recognition*. Minneapolis: University of Minnesota Press.

Osborne, Kayla (2018). Thirteen people arrested at Lakesland chicken farm protest.' *Campbelltown MacArthur Advertiser*, 22 June, https://bit.ly/2GALgXm.

Pachirat, Timothy (2011). *Every twelve seconds*. New Haven: Yale University Press.

Pachirat, Timothy (2012). Interviewed by James McWilliams. *The Atlantic*, 5 June, https://bit.ly/3njWgJC.

Panksepp, Jaak (1998). *Affective neuroscience*. Oxford: Oxford University Press.

Panksepp, Jaak (2010). Interviewed by Ginger Campbell. *Brain science podcast*, 13 January, http://brainsciencepodcast.com/.

Panksepp, Jaak (2011). Cross-species affective neuroscience decoding of the primal affective experiences of humans and related animals. *PLOS ONE* 6(8): e21236, doi:10.1371/journal.pone.0021236.

Panksepp, Jaak, and Lucy Biven (2012). *The archaeology of mind: neuroevolutionary origins of human emotions*. New York: W.W. Norton.

Parkes, Colin M. (1971). Psycho-social transitions: a field study. *Social science and medicine* 5: 101–115.

Parkes, Colin M. (1972). *Bereavement: studies of grief in adult life*. New York: International Universities Press.

Parkes, Colin M. (2009 [2006]). *Love and loss: the roots of grief and its complications.* London: Routledge.

Parkes, Colin Murray, Pittu Laungani, and Bill Young, eds., (1997). *Death and bereavement across cultures.* London: Routledge.

Payne, Peter, Peter A. Levine, and Mardi A. Crane-Godreau (2015). Somatic experiencing: using interoception and proprioception as core elements of trauma recovery. *Frontiers in psychology* 6(art.no.93), https://doi.org/10.3389/fpsyg.2015.00093.

Peirce, J. W., A.E. Leigh, and K.M. Kendrick (2000). Configurational coding, familiarity and the right hemisphere advantage for face recognition in sheep. *Neuropsychologia* 38(4): 475–483.

Peña-Guzmán, David M. (2017). Can nonhuman animals commit suicide? *Animal sentience* 20(1): https://bit.ly/2Gjfqyn.

Phelps, Elizabeth A. (2004). Human emotion and memory: interactions of the amygdala and hippocampal complex. *Current opinion in neurobiology* 14: 198–202.

Philo, Chris (1998). Animals, geography, and the city: notes on inclusions and exclusions. In *Animal geographies: place, politics, and identity in the nature-culture borderlands.* Jennifer Wolch and Jody Emel, eds., London, New York: Verso.

Pierce, Jessica (2013). The dying animal. *Journal of bioethical inquiry* 10(4): 469–478.

Polan, H. Jonathan, and Myron A. Hofer (2016). Psychobiological origins of infant attachment and its role in development. In *Handbook of attachment: theory, research, and clinical applications.* Jude Cassidy and Phillip R. Shaver, eds., 3rd edition, 117–132. New York: The Guilford Press.

Pollan, Michael (2013). The intelligent plant: scientists debate a new way of understanding flora. *The New Yorker*, 23 & 30 December, https://bit.ly/2IIwS05.

Potts, Annie (2012). *Chicken.* London: Reaktion Books.

Potts, Annie, ed. (2016). *Meat culture.* Brill.

Poulton, Edward Bagnall (1890). *The colours of animals, their meaning and use, especially considered in the case of insects.* New York: D. Appleton and Company.

Preti, Antonio (2007). Suicide among animals: a review of evidence. *Psychological reports* 101: 831–848.

Promey, Sally M., ed. (2014). *Sensational religion: sensory cultures in material practice.* New Haven: Yale University Press.

Proust, Marcel (2014 [1913]). *Remembrance of things past [In search of lost time].* Translated by C.K. Scott Moncrieff. Adelaide: University of Adelaide e-book.

Provenza, F.D. (1995). Origin of food preferences in Herbivores. National wildlife research center repellents conference, paper 29, https://bit.ly/3ktcA8g.

Prus, Adam J., John R. James, and John A. Rosecrans (2009). Chapter 4: Conditioned place preference. In *Methods of behavior analysis in neuroscience*. J.J. Buccafusco, ed., 2nd edition. Boca Ranton (FL): CRC Press/ Taylor & Francis, https://bit.ly/38T7zU6.

Querne, L., F. Eustache, and S. Faure (2000). Interhemispheric inhibition, intrahemispheric activation, and lexical capacities of the right hemisphere: a tachistoscopic, divided visual-field study in normal subjects. *Brain and language* 74(2): 171–190.

Ramachandran, Vilayanur S. (2006). Mirror neurons and the brain in the vat. *Edge*, 1 October, https://bit.ly/3lAFi8B.

Rando, Therese A. (1997). Vicarious bereavement. In *Death and the quest for meaning*. Stephen Strack, ed., 257–274. Northvale: Jason Aronson.

Raum, Otto Friedrich (1940). *Chaga childhood*. London: Oxford University Press.

Regan, Tom (2004). Giving voice to animal rights. Interview, *Satya*, August, https://bit.ly/35vPzND.

Reggente, Melissa A.L.V., Elena Papale, Niall McGinty, Lavinia Eddy, Giuseppe Andrea de Lucia, and Chiara Giulia Bertulli (2018). Social relationships and death-related behaviour in aquatic mammals: a systemic review. *Philosophical transactions of the Royal Society B* 373(1754): 20170260.

Reinert, Duane F., and Carla E. Edwards (2009). Attachment theory, child mistreatment, and religiosity. *Psychology of religion and spirituality* 1(1): 25–34.

Reynolds, Pamela (1991). *Dance civet cat: child labour in the Zambezi valley*. Athens, OH: Ohio University Press.

Richards, Emma, Tania Signal, and Nik Taylor (2013). A different cut? Comparing attitudes toward animals and propensity for aggression within two primary industry cohorts – farmers and meatworkers. *Society & animals* 21(4): 395–413.

Rilke, Rainer M. (2009 [1923]). The first elegy. *Duino elegies & the sonnets to Orpheus*. Edited and translated by S. Mitchell. New York: Vintage International.

Roberts, William A. (2007). Mental time travel: animals anticipate the future. *Current biology* 17(11): 418–420.

Rockett, Ben, and Sam Carr (2014). Animals and attachment theory. *Society & animals* 22(4): 415–433.

Rogan, Ruth, Moira O'Connor, and Pierre Horwitz (2005). Nowhere to hide: awareness and perception of environmental change, and their influence on relationships with place. *Journal of environmental psychology* 25(2): 147–158.

Works cited

Rogers, Chris (2009). What became of Romania's neglected orphans? *BBC News*, 22 December, https://bbc.in/3f0ZRsl.

Rohr, Richard (2008). Becoming stillness. Public lecture, Norwich Cathedral, https://bit.ly/3f0ARBz.

Rosenblatt, Paul C. (1997). Grief in small-scale societies. In *Death and bereavement across cultures*. Colin Murray Parkes, Pittu Laungani and Bill Young, eds., 27–51. London: Routledge.

Ross, Heather E., Charlene D. Cole, Yoland Smith, Inga D. Neumann, Rainer Landgraf, Anne Z. Murphy, and Larry J. Young (2009). Characterization of the oxytocin system regulating affiliative behaviour in female prairie voles. *Neuroscience* 162(4): 892–903, https://bit.ly/2IEidUd.

Rossetti, Christina (1896). To what purpose is this waste? *New poems*. London: Macmillan and Co.

Rostila, Mikael, Jan Saarela, and Ichiro Kawachi (2013). Suicide following the death of a sibling: a nationwide follow-up study from Sweden. *BMJ open* 3: e002618.

Russock, Howard I. (1999). Filial social bond formation in fry of maternal mouthbrooding tilapia (pisces: cichlidae): a comparative study. *Behaviour* 136(5): 567–594.

Ryder, Richard (2001). *Painism: a modern morality*. Centaur.

Safina, Carl (2015). *Beyond words: what animals think and feel*. New York: Henry Holt and Company.

Saler, Benson (2009). Anthropomorphism and animism: on Stewart E. Guthrie, *Faces in the clouds* (1993). In *Contemporary theories of religion: a critical companion*. Michael Stausberg, ed., 39–52. London and New York: Routledge.

Samuel, Geoffrey (2008). *The origins of yoga and tantra: Indic religions to the thirteenth century*. Cambridge: Cambridge University Press.

Sangrigoli, Sandy, Christophe Pallier, Anne Marie Argenti, Valerie A. Ventureyra, and Scania de Schonen (2005). Reversibility of the other-race effect in face recognition during childhood. *Psychological science* 16(6): 440–444.

Sapolsky, Robert M. (2004). Mothering style and methylation. *Nature neuroscience* 7(8): 791–792.

Sargent, Carolyn (1984). Between death and shame: dimensions of pain in Bariba culture. *Social science and medicine* 19(12): 1299–1304.

Scannell, Leila, and Robert Gifford (2010). Defining place attachment: a tripartite organizing framework. *Journal of environmental psychology* 30(1): 1–10, https://doi.org/10.1016/j.jenvp.2009.09.006.

Schaefer, Donovan O. (2012). Do animals have religion? Interdisciplinary perspectives on religion and embodiment. *Anthrozoös* 25: S173–189.

Schaefer, Donovan O. (2015). *Religious affects: animality, evolution, and power.* Durham, London: Duke University Press.

Schaffer, H. Rudolph, and Peggy E. Emerson (1964). The development of social attachments in infancy. *Monographs of the society for research in child development* 29(3).

Scheper-Hughes, Nancy (1984). Infant mortality and infant care: cultural and economic constraints on nurturing in northeast Brazil. *Social sciences and medicine* 19(5): 535–546.

Scheper-Hughes, Nancy (1989). Death without weeping: has poverty ravaged mother love in the shantytowns of Brazil? *Natural history* 10: 8–16.

Scheper-Hughes, Nancy (1992). *Death without weeping: the violence of everyday life in Brazil.* Berkeley: University of California Press.

Scheper-Hughes, Nancy (2013). No more angel-babies on the Alto. *Berkeley review of Latin American studies*, Spring: 25–31.

Schichowski C., E. Moors, and M. Gauly (2008). Effects of weaning lambs in two stages or by abrupt separation on their behaviour and growth rate. *Journal of animal science* 68(1): 220–225.

Schore, Allan N. (1982). Foreword. In *Attachment and loss: volume I: attachment* by John Bowlby. 2nd edition, xi–xxi. New York: Basic Books.

Schore, Allan N. (2005). Attachment, affect regulation, and the developing right brain: linking developmental neuroscience to pediatrics. *Pediatrics in review* 26(6): 204–217.

Schore, Allan N. (2011). The right brain implicit self lies at the core of psychoanalysis. *Psychoanalytic dialogues* 21: 75–100.

Schore, Allan N. (2012). *The science of the art of psychotherapy.* New York: W.W. Norton & Company.

Scott, Mark S.M. (2009). Journeys in grief: theorizing mourning rituals. *ARC, The journals of the faculty of religious studies, McGill University* 37: 79–89.

Segal, Z.V., S. Kennedy, M. Gemar, K. Hood, R. Pedersen, and T. Bruis (2006). Cognitive reactivity to sad mood provocation and the prediction of depressive relapse. *Archives of general psychiatry* 63(7): 749–755.

Shaefer, Jodi (1999). When an infant dies: cross-cultural expressions of grief and loss. *Bulletin: a publication of the national fetal-infant mortality review program*, HRSA, U.S. Department of Health & Human Services Public Health Service.

Shahina, K.K. (2010). Mother, shall I put you to sleep? *Tehelka magazin* 7(46), 20 November, https://bit.ly/36Dnwv8.

Sheridan, Margaret A., Nathan A. Fox, Charles H. Zeanah, Katie A. McLaughlin, and Charles A. Nelson III. (2012). Variation in neural development as a result of exposure to institutionalization early in childhood. *Proceedings of the*

National Academy of Sciences of the USA 109(32): 12927–12932, https://doi.org/10.1073/pnas.1200041109.

Shiota, Michelle N., Dacher Keltner, and Amanda Mossman (2007). The nature of awe: elicitors, appraisals, and effects on self-concept. *Cognition & emotion* 21(5): 944–963.

Shonkoff, Jack P., and Deborah A. Phillips, eds., (2000). *From neurons to neighbourhoods: the science of early childhood development.* Washington D.C.: National Academy Press.

Siebert, Charles (2009). Watching whales watching us. *New York Times*, 8 July, https://nyti.ms/3nt773i.

Siegel, Daniel J. (2011 [2010]). *Mindsight: the new science of personal transformation.* New York: Bantam Books Trade Paperbacks.

Silvennoinen, Aino (2017). Changing the face of animal activism. *Artefact*, 5 December, https://bit.ly/36HGY9Q.

Singer, Peter (1993). *How are we to live? Ethics in an age of self-interest.* New York: Prometheus Books.

Slater, Lauren (2004). Monkey love. *The Boston Globe*, 21 March, https://bit.ly/3nsmUzk.

Slobodchikoff, Con (2012). *Chasing Doctor Dolittle.* St Martin's Press.

Smith-Conway, Erin R., Helen J Chenery, Anthony J. Angwin, and David A. Copland (2012). A dual task priming investigation of right hemisphere inhibition for people with left hemisphere lesions. *Behavioral and brain functions* 8(14), https://doi.org/10.1186/1744-9081-8-14.

Snell, Tristan L., and Janette G. Simmonds (2012). 'Being in that environment can be very therapeutic': spiritual experiences in nature. *Ecopsychology* 4(4): 326–336.

Snyder, Allan, Terry Bossomaier, and D. John Mitchell (2004). Concept formation: 'object' attributes dynamically inhibited from conscious awareness. *Journal of integrative neuroscience* 3(1): 31–46.

Sobel, David (1990). A place in the world: adults' memories of childhood's special places. *Children's environments quarterly* 7(4): 5–12.

Spitz, René A. (1945). Hospitalism – an inquiry into the genesis of psychiatric conditions in early childhood. *Psychoanalytic study of the child* 1: 53–74.

Stanescu James (2012). Species trouble: Judith Butler, mourning, and the precarious lives of animals. *Hypatia* 27(3): 567–582.

Stedman, R. C. (2003). Is it really just a social construction? The contribution of the physical environment to sense of place. *Society and natural resources* 16(8): 671–685.

Stevenson, I. Neil (1977). *Colerina*: reactions to emotional stress in the Peruvian Andes. *Social science and medicine* 11(5): 303–307.

Stewart, Cameron (2013). Naming taboo often ignored in breaking news. *The Australian*, 13 July, https://bit.ly/33xQex7.

Striniste, Nancy A., and Robin C. Moore (1989). Early childhood outdoors: a literature review related to the design of childcare environments. *Children's environments quarterly* 6(4): 25–31.

Sugiyama, Yukimaru, Hiroyuki Kurita, Takeshi Matsui, Satoshi Kimoto, and Tadatoshi Shimomura (2009). Carrying of dead infants by Japanese macaque (*Macaca Fuscata*) mothers. *Anthropological science* 117(2): 113–119.

Sullender, Scott R. (2010). Vicarious grieving and the media. *Pastoral psychology* 59(2): 191–200.

Sunda, Mike (2015). Japan's hidden caste of untouchables. *BBC News*, 23 October, https://bbc.in/2UxDrp0.

Suomi, Stephen J. (2016). Attachment in rhesus monkeys. In *Handbook of attachment: theory, research, and clinical applications*. Jude Cassidy and Phillip R. Shaver, eds., 3rd edition, 133–154. New York: The Guilford Press.

Takahashi, Tetsumi, Haruki Ochi, Masanori Kohda, and Michio Hori (2012). Invisible pair bonds detected by molecular analyses. *Biology letters* 8(3): 355–357, doi: 10.1098/rsbl.2011.1006.

Taylor, Nik, and Heather Fraser (2019). *Companion animals and domestic violence: Rescuing you, rescuing me*. London: Palgrave.

The farmer's magazine, 1849, XIX(1): 142, https://bit.ly/33yArOK.

Tibbets, Elizabeth A., Ellery Wong, and Sarah Bonello. Wasps use social eavesdropping to learn about individual rivals. *Current biology* 30(15): 3007–10, doi: https://doi.org/10.1016/j.cub.2020.05.053.

Tiesman, Hope M., Srinivas Konda, Dan Hartley, Cammie Chaumont Menéndez, Marilyn Ridenour, and Scott Hendricks (2015). Suicide in U.S. workplaces, 2003-2010: a comparison with non-workplace suicides. *American journal of preventive medicine* 48(6): 674–682, doi:10.1016/j.amepre.2014.12.011.

Tolstoy, Leo (2009 [1892]). Introduction – the first step. In *The ethics of diet: an anthology of vegetarian thought*. Howard Williams, abridged version. Guildford: White Crows Books.

Torta D.M. and F. Cauda (2011). Different functions in the cingulate cortex, a meta-analytic connectivity modelling study. *NeuroImage* 56(4): 2157–2172.

Townend, Christine (2017). *A life for animals*. Sydney: Sydney University Press.

Ulrich, R.S. (1984). View through a window may influence recovery from surgery. *Science* 224(4647): 420–421.

UPC (2018). Forced molting. 3 February (update), https://bit.ly/3px9cNh

Valdesolo, Piercarlo, and Jesse Graham (2014). Awe, uncertainty and agency detection. *Psychological science* 25(1): 170–178.

Vallortigara, Giorgio, and Lesley J. Rogers (2005). Survival with an asymmetrical brain: advantages and disadvantages of cerebral lateralization. *Behavioral and brain sciences* 28: 575–633.

Vallortigara, Giorgio, and R.J. Andrew (1994). Differential involvement of right and left hemisphere in individual recognition in the domestic chick. *Behavioural processes* 33: 41–58.

Vallortigara, Giorgio, Allan Snyder, Gisela Kaplan, Patrick Bateson, Nicola S. Clayton, and Lesley J. Rogers (2008). Are animals autistic savants?' *PLOS biology* 6(2): 0208–0214, https://doi.org/10.1371/journal.pbio.0060042.

Van der Geest, Sjaak (2004). Dying peacefully: considering good death and bad death in Kwahu Tafo, Ghana. *Social science and medicine* 58(5): 899–911.

van Dooren, Thom (2010). Pain of extinction: the death of a vulture. *Cultural studies review* 16(2): 271–289.

Vanlancker-Sidtis, Diana (2004). When only the right hemisphere is left: studies in language and communication. *Brain and language* 91: 199–211.

Venkatraman, Anand, Brian L. Edlow, and Mary Helen Immordino-Yang (2017). The brainstem in emotion: a review. *Frontiers in neuroanatomy* 11(15), doi: 10.3389/fnana.2017.00015.

Victorian Government DPI (2007). Code of accepted farming practice for the welfare of sheep. Revision 2. *Victorian Government Gazette,* 21 June, https://bit.ly/32LmgF0

Villanueva, Gonzalo (2018). *A transnational history of the Australian animal movement, 1970–2015.* Palgrave Macmillan.

Vingerhoets, Guy, Celine Berckmoes, and Nathalie Stroobant (2003). Cerebral hemodynamics during discrimination of prosodic and semantic emotion in speech studied by transcranial doppler ultrasonography. *Neuropsychology* 17(1): 93–99.

Vizcaíno, Elena, M.A. de Andrés, José Alberto Murillo Murillo, Carlos Piñeiro, and María Aparicio (2012). Do we underestimate the lameness in the sows? *Farm data analysis,* 9 July, https://bit.ly/3lfN2Mu.

Wahlberg, David (2015). Controversial UW-Madison monkey study won't remove newborns from mothers. *Wisconsin state journal,* 13 March, https://bit.ly/34nOKVm.

Ward, Michael (1997). Reasons for decision of Magistrate Ward. Magistrates Court Canberra A.C.T.

Watanabe, Zenjiro (2005). The meat-eating culture of Japan at the beginning of Westernization. *Food culture* 9: 2–8, https://bit.ly/2ID4a0X.

Watson, Claire F.I., and Tetsuro Matsuzawa (2018). Behaviour of nonhuman primate mothers toward their dead infants: uncovering mechanisms. *Philosophical transactions of the Royal Society B* 373(1754): 20170261.

Watt, Stuart (1998). Seeing this as people: anthropomorphism and common-sense psychology. Ph.D. thesis, The Open University.

Weary, Daniel M., and David Fraser (2009). Social and reproductive behaviour. In *The ethology of domestic animals: an introductory text*. Per Jensen, ed., 2nd edition, 73–84. Wallingford (UK): CABI.

Weaver, Ian C.G., Nadia Cervoni, Frances A. Champagne, Ana C. D'Alessio, Shakti Sharma, Jonathan R. Seckl, Sergiy Dymov, Moshe Szyf, and Michael J. Meaney (2004). Epigenetic programming by maternal behaviour. *Nature neuroscience* 7(8): 847–854.

Webster, Donna, and Arie W. Kruglanski (1994). Individual differences in need for cognitive closure. *Journal of personality and social psychology* 67(6): 1049–1062.

Weiss, Robert S. (1974). *Loneliness: the experience of emotional and social isolation*. Cambridge, MA: MIT Press.

Weiss, Robert S. (1977). *Marital separation*. New York: Basic Books.

West, Meredith J., Andrew P. King, and David J. White (2003). The case for developmental ecology. *Animal behaviour* 66: 617–622.

White, Geoffrey M., Tobias Uller, and Erik Wapstra (2009). Family conflict and the evolution of sociality in reptiles. *Behavioural ecology* 20(2): 245–250, https://doi.org/10.1093/beheco/arp015.

Willett, Cynthia (2013). Water and wing give wonder: trans-species cosmopolitanism. *PhaenEx* 8(2): 185–208.

Willett, Cynthia (2014). *Interspecies ethics*. New York: Columbia University Press.

Wilson, Edward O. (2009). The ant cemetery. *World science festival*, interviewed by Josh Zepps, https://bit.ly/3kAuINi.

Wright, Anthony (n.d.). Chapter 5: Limbic system: hippocampus. Chapter 6: Limbic system: amygdala. *Neuroscience online: an electronic textbook for the neurosciences*, Department of Neurobiology and Anatomy, University of Texas, https://bit.ly/3nyzCwx and https://bit.ly/3f1UFo1.

Yeates, J.W., and D.C.J. Main (2008). Assessment of positive welfare: a review. *The veterinary journal* 175(3): 293–300.

Yoerg, Sonja (1992). Mentalist imputations. *Science* 258(5083): 830–831.

Yuasa, Yasuo (1987). *The body: toward an eastern mind body theory*. Edited by Thomas P. Kasulis, translated by Nagatomo Shigenori and T.P. Kasulis. New York: New York State University Press.

Yun, Jinhyeon, and Anna Valros (2015). Benefits of prepartum nest-building behaviour on parturition and lactation in sows – a review. *Asian-Australasian journal of animal sciences* 28(11): 1519–1524, doi: 10.5713/ajas.15.0174.

Zhao, Andong, Hua Qin, and Xiaobing Fu (2016). What determines the regenerative capacity in animals? *BioScience* 66(9): 735–746, https://doi.org/10.1093/biosci/biw079.

Zhu, Ying, Li Zhang, Jin Fan, and Shihui Han (2007). Neural basis of cultural influence on self-representation. *NeuroImage* 34: 1310–1316.

Zilcha-Mano, Sigal, Mario Mikulincer, and Phillip R. Shaver (2012). Pets as safe havens and secure bases: the moderating role of pet attachments orientations. *Journal of research in personality* 46(5): 571–580.

Zuidhof, Martin, B.L. Schneider, V.L. Carney, D.R. Korver, and F.E. Robinson (2014). Growth, efficiency, and yield of commercial broilers from 1957, 1978, and 2005. *Poultry science* 93(12): 1–13.

Acknowledgements

This work is based on my PhD dissertation, which I undertook part-time under the supervision of Jay Johnston at the University of Sydney. My sincere gratitude goes to Jay and the entire department for their guidance, kindness and support over the years. I would like to thank the external examiners, Cynthia Willett, Kocku von Stuckrad and Peta Tait, who reacted very positively to the thesis and encouraged publication. Over the years multiple ideas from this book were published as journal articles and/or book chapters and I would like to extend my thanks to the editors and anonymous peer-reviewers who were generally very encouraging and whose input was paramount for further development of the ideas. Many – perhaps *most* – of these ideas emerged and matured with the help of human and nonhuman animals in the field; I would like to thank all those nonhuman animals who opened up and let me into their world, and the inspiring humans I have worked with over the years and/or who have shared their thoughts and experiences with me, particularly sanctuary providers, rescuers/carers and activists/advocates, many of whom have become close friends, an honour beyond words. In the later developmental stages, this work benefitted substantially from the input of Lori Marino, who kindly agreed to a critical reading of selected sections of the manuscript, and of the two anonymous reviewers. The publication would not have been possible without the enthusiastic support and professionalism of the

Sydney University Press team: the editors, Agata Mrva-Montoya, Melissa Boyde and Fiona Probyn-Rapsey; publishing manager Denise O'Dea; copyeditor Jo Butler; production officer Nathan Grice; and designer Miguel Yamin.

My deepest gratitude to and loving appreciation for David, who has always stood by me, in the good times and the bad times, as we had promised each other, and, of course, to Charlie, our recently deceased canine companion, and to Orpheus-Pumpkin, Henry, Jonathan and Jason, our family's ovine component.

Index